Spyridon Korres

On-Line Topographic Measurements
of Lubricated Metallic Sliding Surfaces

A case study of lubricated buried interfaces

**Schriftenreihe
des Instituts für Angewandte Materialien**
Band 20

Karlsruher Institut für Technologie (KIT)
Institut für Angewandte Materialien (IAM)

Eine Übersicht über alle bisher in dieser Schriftenreihe erschienenen Bände
finden Sie am Ende des Buches.

On-Line Topographic Measurements of Lubricated Metallic Sliding Surfaces

A case study of lubricated buried interfaces

by
Spyridon Korres

Dissertation, Karlsruher Institut für Technologie (KIT)
Fakultät für Maschinenbau
Tag der mündlichen Prüfung: 30. August 2012

Impressum

Karlsruher Institut für Technologie (KIT)
KIT Scientific Publishing
Straße am Forum 2
D-76131 Karlsruhe
www.ksp.kit.edu

KIT – Universität des Landes Baden-Württemberg und
nationales Forschungszentrum in der Helmholtz-Gemeinschaft

KIT Scientific Publishing 2013
Print on Demand

ISSN 2192-9963
ISBN 978-3-7315-0017-9

On-Line Topographic Measurements of Lubricated Metallic Sliding Surfaces

Zur Erlangung des akademischen Grades

Doktor der Ingenieurwissenschaften

der Fakultät für Maschinenbau

Karlsruher Institut für Technologie (KIT),

genehmigte

Dissertation

von

M.Sc. Spyridon Korres

aus Athen, Griechenland

Tag der mündlichen Prüfung: 30. August 2012
Hauptreferent: PD Dr. rer. nat. Martin Dienwiebel
Korreferent: Prof. Dr. rer. nat. Oliver Kraft

To my family.
Στην οικογένειά μου.

Abstract

A great part of the applications that involve sliding surfaces are strongly dependent on the running-in of the tribo-systems. Lately, more and more industrial research is focused on the improvement of running-in, in order to achieve reduced wear rates and minimal energy loss due to friction. In this process, the friction coefficient and environmental parameters are monitored throughout the tribological load. Post load experiments include chemical analysis of the surface, topography examination and wear measurement. The radionuclide on-line wear measurement was a major breakthrough in modern tribology, as it allows researchers to correlate the wear rates with the friction coefficient during such an experiment. Nevertheless, one of the most critical parameters for the understanding of running-in, the topography of the sliding surfaces, can only be determined at the end of every test. The reason why the topographical evolution during the running-in of lubricated sliding surfaces remains unclear lies in the difficulty of *in-situ* measurements in such experimental layouts. In this work, a state-of-the-art tribometer was constructed to investigate the dynamic behavior of sliding metallic surfaces under lubrication. The basis of the system is a positioning system with sub-micron precision and a custom-made force sensor that can measure x,y and z forces simultaneously. The topography is measured with a 3D holography microscope at a maximum frequency of 15 fps as well as an atomic force microscope (AFM). The samples and the objective lens of the microscope are immersed in poly alpha olefin (PAO). A radionuclide technique apparatus measures wear on-line. The entire instrument is controlled with custom built software which is also used for the processing of the results. Tests performed with steel and iron samples demonstrate stable function of the equipment. Taking advantage of the features of this instrument, two blocks of experiments were performed. The first block was performed with flat steel pins against flat Cu samples in a linear reciprocating motion, as well as in a uni-directional manner. The results suggest that mechanical mixing and material transfer lead to an unstable lamellar formation near the surface. Still, delamination cannot be excluded as a possible mechanism in the examined system. The second block was performed with ruby spheres against flat Cu samples, again in a linear reciprocating motion. According to the results, an approximation technique is proposed to precisely separate plowing from shear terms of the friction force. This is possible by measuring the widening rate of plowing tracks. The theory of Bowden, Moore and Tabor was used to explain the results.

Kurzfassung

Ein großer Teil der Anwendungen, die Gleitflächen beinhalten, hängen stark vom Einlaufverhalten des Tribosystems ab. Inzwischen konzentriert sich die industrielle Forschung immer mehr auf die Verbesserung des Einlaufverhaltens, um die Verschleißraten und den Energieverlust durch Reibung zu minimieren. Bei diesen Untersuchungen werden die Reibkoeffizienten und die Umgebungsbedingungen während der gesamten tribologischen Belastungsphase aufgezeichnet. Die anschließenden Untersuchungen beinhalten die chemische Analyse der Oberfläche, die Topographieanalyse und die Verschleißmessung. Die Radionuklid-on-line-Verschleißmessung war dabei ein großer Durchbruch der modernen Tribologie, da sie es den Forschern ermöglicht, die Verschleißraten mit den Reibkoeffizienten während des Experiments zu korrelieren. Nichtsdestotrotz kann die resultierende Topographie der Gleitflächen, welche einen der kritischsten Parameter für das Verständnis des Einlaufprozesses darstellt, stets erst am Ende eines jeden Versuchs bestimmt werden. Der Grund, weshalb die Topographieentwicklung während des Einlaufprozesses von geschmierten Gleitflächen unklar bleibt, liegt in der Schwierigkeit der *in-situ*-Messung dieser experimentellen Ausführung. Im Rahmen der vorliegenden Arbeit wurde ein hochmodernes Tribometer konstruiert, um das dynamische Verhalten von geschmierten, metallischen Gleitflächen zu untersuchen. Die Grundlage des Aufbaus ist ein submikrometer-präzises Positioniersystem und ein selbst gefertigter Kraftsensor, der gleichzeitig in alle drei Raumrichtungen die Kräfte messen kann. Die Topographie wird sowohl mit einem 3D-Holographie-Mikroskop mit einer maximalen Bildrate von 15 Aufnahmen pro Sekunde wie auch mit einem Rasterkraftmikroskop (AFM) gemessen. Die Proben und das Objektiv des Mikroskops sind in Polyalphaolefin (PAO) getaucht. Ein Radionuklidtechnik-Gerät misst den Verschleiß online.
Der gesamte Versuchsaufbau wird mit einer selbstgeschriebenen Software gesteuert, die auch für die Datenverarbeitung genutzt wird. Versuche mit

Stahl- und Eisenproben zeigten die stabile Funktion des Aufbaus. Die Vorzüge des Tribometers wurden genutzt, um zwei Versuchsblöcke durchzuführen. Der erste Block wurde mit flachen Stahlstiften durchgeführt, die linear-reversierend wie auch in einsinniger Bewegungsrichtung auf flachen Kupferproben gleiten. Die Ergebnisse deuten darauf hin, dass eine mechanische Durchmischung und ein Materialübertrag zu einer instabilen Lamellenbildung nahe der Oberfläche führen. Jedoch kann Delamination als möglicher Verschleißmechanismus im untersuchten System nicht ausgeschlossen werden. Der zweite Versuchsblock wurde mit linear- reversierenden Rubinkugeln auf flachen Kupferproben durchgeführt. Basierend auf den Ergebnissen wird eine Näherungsmethode vorgeschlagen, um die Pflüg- und Scherbeiträge der Reibkraft präzise zu trennen. Dies wird erst durch die Messung der Verbreiterungsrate der Pflügspur möglich. Die Theorie von Bowden, Moore und Tabor wurde zur Erklärung der Ergebnisse herangezogen.

Acknowledgments

The presented work was completed between March 2008 and December 2011 at the Karlsruher Institut für Technologie (KIT) and at the Fraunhofer Institut für Werkstoff- mechanik (IWM)/MikrotribologieCentrum (μTC), as part of project funded by the German Science Foundation (DFG) under grant number DI 1494/1-1.

Firt of all I would like to thank Dr. Martin Dienwiebel, who gave me the opportunity to work in his research group and introduced me not only to the field of Tribology, but also to some of the most important researchers in the tribological society. In the same scope, I would like to thank Prof. Dr. Matthias Scherge and Prof. Dr. Gumbsch for their support.

This work was made possible thanks to the support that I received from the technical staff of the following companies: TETRA Gesellschaft für Sensorik, Robotik und Automation mbH, Lyncée Tec SA and Zyklotron AG. I am most grateful for their assistance.

For the results presented in this work, I acknowledge the contribution of my colleagues in Karlsruhe and Freiburg. Special thanks go to Tim Feser, Diego Marchetto, Eberhatd Nold, Anson Santoso Wong and Peter Frühberger. I would also like to express my gratitude for Thi-Bach-Hue Quan's cheerfulness every morning at work. I have shared great moments at work with all these people. For the rest of my time out of the office/lab, I'd like to thank my friends.

I would also like to thank those who helped me complete the final draft of my thesis (Nick Verriotis, Alexander Renz and Tobias Hoppe).

Above all, I would like to thank my family, who have been there for me throughout it all.

Contents

1. Introduction

Tribology is a scientific and engineering field that combines elements of solid mechanics, fluid mechanics, materials science and chemistry. It is known that our ancestors already understood the importance of friction in 1800 BC in ancient Egypt, when constructing vehicles for the Pharaoh [1]. Leonardo da Vinci was probably one of the first scientists to systematically study friction. He reported that the areas of contact do not have an effect on friction and that if the load is doubled, the friction will also be doubled. He even made sketches of antifriction bearings. In 1966 Jost wrote a report that stressed the impact of friction and wear on economy [2]. It was in this landmark report that the term "tribology" was introduced, a word that comes from the Greek word "τρίβω", which means to rub. The field of tribology has attracted a lot of attention since this time, especially in countries with increased industrial activity [3].

In recent years, phenomena related to friction and wear have received strong interest in academia and industry. A main reason is the emerging potential of achieving higher efficiency in various processes by reducing friction and improving the reliability of parts via wear reduction. The environment, surface and subsurface structure of bodies after being subjected to tribological loads have been studied extensively. Most results of post mortem examinations indicate, that during the application of load, complex phenomena occur.

Among the current trends in tribology, *in-situ* measurements begin to shine, thanks to various breakthroughs in modern instrumentation. Several modern *in-situ* methods aiming at elucidating the dynamics of these very phenomena are referenced in the following chapters. An important observation when evaluating such methods is that monitoring a system during a tribological experiment usually poses several limitations with respect to the conditions of the experiment. Examples of such limitations can be performing sliding tests with transparent bodies, in dry environments, under vacuum or inert gas etc. (see Chapter 2). The majority of advanced industrial tribo-systems involve lubricated metallic systems.

The main aim of this work is to develop a method that allows for in-situ investigation of lubricated metallic sliding surfaces, focusing on topography changes which can be correlated with wear, subsurface restructuring and fluctuations of the friction coefficient. The argumentation in the discussion of the results is based on the roots of tribology's theory, so this first chapter provides a brief explanation of the very fundamentals of tribology. Furthermore, a novel experimental technique is introduced in this work. Therefore, an overview of the most important tribological testing methods is given, in order to understand the role and position of the technique in comparison to the existing experimental paths.

1.1. The Nature of Friction

Some of the basic concepts in tribology begin with Amonton's contribution. Part of his work was the empirical discovery of two important laws of dry friction:

1. The friction force is independent of the apparent contact area.

2. The friction force is directly proportional to the applied normal force.

$$F_F = \mu F_N,\tag{1.1}$$

μ being the friction coefficient.

In addition to the above, Coulomb introduced the third law of dry friction, according to which the kinetic friction is independent of the sliding velocity.

Coulomb and Amonton claimed that the real contact area does not depend on the apparent contact area or the sliding speed. It is however proportional to the normal force.

The mechanisms of metallic friction were further studied by Bowden, Moore and Tabor, with a particular emphasis on the role of shearing and plowing in dry, wearless sliding [4]. According to their consideration, metallic junctions larger than molecular dimensions are formed and sheared during sliding. Furthermore, provided that one of the two surfaces in contact is harder, the asperities of the harder surface will plow out a certain volume of the counterface.

1.1.1. Plastic junction model (adhesion model)

According to Bowden, Moore and Tabor, the force F_S required to shear the metallic junctions formed at the points of intimate contact between two metals is given by the following equation,

$$F_S = A_r s, \tag{1.2}$$

where A_r is the real contact area and s is the force per unit area which, acting in a direction tangential to the interface, is required to shear the junctions. The force per unit area s was later replaced by σ, the shear strength, which in the adhesive theory was related to tensile experiments, while the symbol that is currently used is usually τ.

The above model is also known as the Bowden and Tabor adhesion model, as it resembles models used in adhesion. This name often leads to the misconception, that friction in their model originates from adhesive forces. All changes of the asperities in this model are assumed to be plastic. The energy loss because of friction is therefore the energy of plastic deformation near the surface.

In common applications, the real contact area of multiple asperity systems is hard to define. In addition, the exact distribution of the normal force, as well as variations in the chemistry of the surfaces and the medium between them cannot be perfectly defined. It is clear, that F_S, as measured in experiments where no wear occurs, is a collective contribution of single asperities coming in contact with the counter-face. When working in ideal conditions of vacuum and a single contact, all parameters mentioned above are easier to determine; however, atomistic approaches, such as the Prandtl-Tomlinson model [5] and MD simulations [6] may be more precise in more complex conditions. Nevertheless, even nowadays, the Bowden-Tabor plastic junction model in sliding friction is commonly used by tribologists, regardless of the scale at which they work.

1.1.2. Plowing friction

Bowden and Tabor are mostly known for their well established adhesive model, although, according to their findings, the contribution of shear is only one of the two components affecting the friction force. The adhesive model only applies to systems without wear. When two soft metallic surfaces slide against each other, plowing is highly likely to occur, even if it is limited to a small scale. In this case an additional term has to be added, in order to consider the resistance encountered when material placed in front of the slider is being plastically deformed. The force F_P required to displace the softer metal from the front of the slider is equal to the cross-section of the grooved track A' multiplied by the mean pressure p' required to displace the metal in the surface [4].

$$F_P = A' p'. \tag{1.3}$$

In this process, an entire volume near the surface is affected. It is therefore expected, that the bulk properties of the softer material greatly influence F_P. Still the pressure p' is not identical with the yield pressure p of indentation, although it is expected to be in the same order of magnitude and is possibly a general function thereof.

It was also proven, that the presence of lubricant cannot influence the plowing term. Despite accepting that a lubricant is not affecting F_F, it can help eliminate friction increases at a later stage of a sliding experiment, when the displaced material accumulates in front of the slider.

1.1.3. From friction forces to friction coefficient

To sum up, the friction force consists of two terms: the shear term that is based on the adhesion model and the plowing term,

$$F_F = F_S + F_P. \tag{1.4}$$

The shear term depends on the real contact area and the nature of the junctions formed at the points of intimate contact, whereas the plowing term is defined by the mechanical bulk properties of the softer body and the amount of material displaced.

The laws of friction give a relation between the friction force and the normal force by multiplying it with the friction coefficient. When a single asperity of the harder body is perfectly spherical and is being indented in a perfectly flat counter-face, it will sink until the normal load is supported by the softer body. The contact area in this case is

$$A_r = \frac{F_N}{p}. \tag{1.5}$$

If plowing is negligible for a tribo-system, then the friction force can be given by substituting the real contact area in equation 1.2 with the result of equation 1.5.

$$F_F = F_N \frac{s}{p}. \tag{1.6}$$

If the asperities are in fully plastic conditions, then p is close to the hardness of the softer material H. After dividing both terms by the normal load, the left hand side of the equation is the friction coefficient, so according to equation 1.1,

$$\mu = \frac{s}{p} \approx \frac{s}{H}.$$ (1.7)

In an ideal system, the plastic deformation should only take place on the softer surface, so s is the shear strength of the softer metal. Furthermore, the surface and the bulk mechanical properties of the two bodies are the same. Based on these assumptions, the Bowden, Moore and Tabor theory connects the friction coefficient to bulk properties of the softer body (the shear strength of the softer metal and the yield pressure of the softer metal or its hardness).

This theory had two shortcomings. The first one was that it did not consider the influence of contaminant films. The second one was, that even in clean environments, the friction coefficients calculated were very low, so the concept of junction growth was introduced. According to this approach, the real contact area determined by the normal load is increased due to the tangential forces. It is noteworthy, that a fully plastic behavior is rather unrealistic in most tribo-systems.

1.2. Contact Area

One of the most important parameters when studying friction between two surfaces in relative motion, as seen above, is the contact area. Most surfaces we encounter in our everyday life have a non-negligible roughness. Polished objects that appear perfectly smooth to the naked eye can reveal amazing complexity when examined with a microscope. Once two such surfaces are pressed together, their structure will only allow a small fraction of what seems to be touching to form the real contact area. The real contact area remains a field of continuous research. Computer simulations have provided scientists with a powerful toolbox to calculate contact areas for a wide variety of systems. This section, however, only contains the very fundamental theoretical approaches to contact mechanics.

The work of Heinrich Hertz formed an important basis for most contact theories [7]. His studies were based on the following assumptions:

1. The strains involved are within the elastic limit.

2. The contact region is much smaller than the radii of curvature and dimensions of the bodies. The bodies are elastic half spaces and the contact area plane has pressures applied normal to it.

3. The surfaces are continuous and non-conforming.

4. The surfaces are frictionless. (This condition is relaxed when dealing with friction.)

A finite contact, called a footprint, is formed in a touching contact between two non-conformal elastic body surfaces of different radii of curvature and Young's moduli. Considering the mechanical properties and the geometry of the components of this footprint, the real contact area can be calculated as follows,

$$A_r = \pi \left(\frac{R^*}{K} \right)^{\frac{2}{3}} F_N^{\frac{2}{3}}, \tag{1.8}$$

with $R^* = \frac{R_1 R_2}{R_1 + R_2}$, R_1 and R_2 being the radii of the two spheres,

$$\frac{1}{K} = \frac{3}{4} \left(\frac{1 - v_1^2}{E_1} + \frac{1 - v_2^2}{E_2} \right),$$

with E_1 and E_2 being the Young's moduli of the two spheres, v_1 and v_2 the Poisson's ratios of the two spheres, and F_N the normal load.

Additional parameters have to be considered when a contact deviates from the above assumptions. Such deviations do not only concern the geometry of the contacts, but also their nature. Adhesion is an important factor that was further studied by Johnson, Kendall, and Roberts (JKR) [8]. Derjaguin-Muller-Toporov's (DMT) [9, 10] model also considered attractive interactions outside the area of contact by introducing Lenard Jones potentials.

Adhesive forces can vary in nature, but the way they affect a tribo-system is always the same. Namely, they add up to an increased normal load. These forces can be categorized as follows:

- Van der Waals forces

- Electrostatic forces

- Capillary forces

- Chemical bonds or acid-base interactions

The work of adhesive forces according to Maugis and Dugdale [11] is

$$\Delta\gamma = \sigma_0 h_0, \tag{1.9}$$

where σ_0 is the maximum force predicted by the Lennard Jones potential and h_0 is the maximum separation obtained by matching the areas under Dugdale and Lennard Jones curves of a contact.

In most applications, perfectly flat or perfectly spherical bodies do not exist. Nevertheless, a surface can be described with a gaussian height distribution and it can be assumed that the peaks are spherical, as described by Greenwood and Williamson [12]. At moderate pressure, the contact area is proportional to the normal load [13]. Plastic deformation may occur at smaller asperities. The probability of this occurring depends on the surface topography and material properties. To quantify this probability, Greenwood and Williamson introduced a dimensionless parameter ψ called plasticity index, which determines if plastic deformation should take place.

$$\psi = \left(\frac{E'}{H}\right) \times \left(\frac{\sigma^*}{r}\right)^{0.5}, \tag{1.10}$$

where E' is the composite Young's modulus, H is the hardness of the deforming surface, σ^* is the standard deviation of the surface peak height distribution and r is the asperity radius. The composite Young's modulus can be calculated from the following equation:

$$\frac{1}{E'} = \frac{1 - v_1^2}{E_1} + \frac{1 - v_2^2}{E_2}.$$

For $\psi < 0.6$ the contact is elastic and for $\psi > 1$ the contact is plastic.

A more modern approach to contacts between two bodies was developed by B.N.J. Persson [14]. His initial work mainly focused on contacts between rubber and hard, rough surfaces, but was further applied to any elastic body in contact with rough, hard surfaces. The random roughness of the harder body was considered on different length scales, describing it as a self-affine fractal distribution, down to a low/distance cut-off, which cannot be smaller than the atomic dimension. In this approach, the rubber does not only undergo elastic deformation at the asperities, locally, but it conforms to larger scale cavities formed by the roughness of the counter-body. Nevertheless, it cannot reach full coverage of the smaller scale surface features of the harder body. Furthermore, the elastic deformation caused by the local pressure at an asperity affects the contact of neighboring asperities.

The theory of B.N.J. Persson predicts a linear increase of the area of contact with the load for small loads. If the length of a macroscopic contact area is L and the length of an asperity in contact is λ, then the contact area depends on the magnification as follows,

$$A(\lambda) \sim (\lambda/L)^{1-J},$$

with J being an exponent related to the fractal dimension.

1.3. Wear

In nature, wear can be found in various forms. Wear, as studied by tribologists, is the progressive loss of surface material due to the relative tangential motion of two bodies in contact. The three classical mechanisms of mechanical wear of metals are: adhesive wear, abrasive wear and fatigue wear.

1.3.1. Adhesive wear

The models that can be used for adhesive wear are based on the work of Archard, originally on abrasive wear, which is also presented in the following section [15]. The main advantage of this model is its simplicity; however, it is usually considered oversimplified when used to describe systems encountered in most applications. Archard considered hemispherical asperities, that have deformed plastically after coming into contact with the counter-face. These asperities experience shear forces as predicted in the adhesive friction model. As the two bodies continue to slide tangentially, shear may occur at the interface if it is weak, or under the surface, if the interface is too strong. In the first case, which is common when an oxide layer covers the surfaces, no wear is expected. In the second case, a fragment of the surface will be plucked away and may be released immediately, or carried further and released after subsequent encounters with asperities of the counter-face. This fragment becomes wear debris.

The area of contact for one of the n hemispherical asperities of radius r is πr^2 and in ideal elastic-plastic conditions with full plastic deformation, the supported load is $\pi r^2 H$ and the worn volume at one asperity when sliding a distance of $2r$ is $2\pi r^3/3$. The total volume loss V for all n asperities when sliding a distance L is

$$V = n\frac{2}{3}\pi r^3 \frac{L}{2r}. \qquad (1.11)$$

Assuming that the contact pressure for plastic deformation at a junction is almost equal to the hardness of the softer material, according to equation 1.5:

$$n\pi r^2 = \frac{F_N}{H}, \qquad (1.12)$$

so equation 1.11 becomes

$$V = \frac{F_N L}{3H}.$$ (1.13)

In nature, adhesive wear can take place in different modes [16], producing particles of sizes that are not always in accordance with the junction areas. In addition, as mentioned above, not all junctions will produce wear debris, as shear can be supported directly at the interface. Finally, various parameters are not considered in this approach. Such factors are wear occurring on the harder body, the subsurface structure and its imperfections, chemical contaminants on the surface, and geometrical variations of the asperities. Due to these inconsistencies an empirical wear coefficient K_{Ad} ($K_{Ad} \ll 1$) was introduced to accommodate these parameters. This coefficient depends on the tribological system, comprising the two bodies in contact and the medium in between. Equation 1.13 now becomes

$$V = K_{Ad} \frac{F_N}{H} L.$$ (1.14)

This means, that wear depends on the sliding distance L. If $K_A d$ remains constant, then the system undergoes steady state wear, whereas if $K_A d$ constantly changes, then running-in wear occurs. In the case of steady state wear, the wear volume is proportional to the sliding time, if the sliding speed is also constant.

1.3.2. Abrasive wear

This mechanism describes severe wear due to deeper penetration, mainly by cutting. Deformation of the softer body is caused by the harder counterface, the third body, or particles trapped between the two. The two first cases are called two-body abrasion, whereas the third one is called three-body abrasion. The geometry used to describe abrasive wear is usually conical, but following the approach described in adhesive wear and incorporating geometrical factors in the empirical wear coefficient for abrasive wear K_{Ab}, the abrasive wear equation is similarly

$$V = K_{Ab} \frac{F_N}{H} L. \tag{1.15}$$

In the above description of abrasive wear, only plastic deformation is considered. In many cases, abrasive wear can be supported by crack propagation, especially in brittle materials. In this case, the wear rate of the brittle material also depends on fracture toughness K_c and Young's Modulus E [17].

1.3.3. Fatigue wear

Fatigue wear is loss of surface material due to the application of cyclic load. The most common type of fatigue wear is in rolling under elastic contact. In this case, the load is not high enough to introduce yield in the contact region, yet in cyclic conditions high-cycle fatigue leads to material failure. In this mechanism, it is very important to consider a large number of parameters affecting the wear behavior of the bodies around the contact area. The subsurface of most materials is highly heterogeneous and contains micro-defects. When load is applied, the presence of grain boundaries, inclusions, voids in polycrystalline materials or even slip planes of single crystals can cause the local stress to increase near the contact, exceeding the yield stress. Work hardening may also occur under the surface until pits start appearing on the surface. The nucleation and propagation of cracks under the surface are strongly dependent on the microstructure of the subsurface; however, they are expected to be found mostly near the points of maximum shear stress. This mechanism is often categorized as macroscopic fatigue wear. It can also occur in sliding conditions, giving rise to significantly smaller pits.

Another mechanism that has to be considered when sliding is that of fatigue wear under plastic contact. In this mechanism, a shallow conformable groove is formed during plowing. Multiple passes of the slider on these grooves cause particles to detach due to plastic burnishing. Plastic flow leads to protrusion of a thin film which gives rise to wear. The number of cycles to failure is usually calculated with modified versions of the Coffin-Manson relationship. For the wear calculation, the coefficient of fatigue wear is given by the following equation,

$$K_F = \frac{3 \cdot 3^{1/2} r_p \mu}{C_s^D \Delta \gamma_s^{1-D}},$$
(1.16)

where C_s is the monotonic effective shear strain, $\Delta \gamma_s$ is the effective shear strain increment per wave pass, D is a constant, and r_p is the ratio of plastic to total work of sliding [18].

These mechanisms will be further discussed in detail in chapter 3.

1.3.4. Corrosive wear

Finally, wear can also be of chemical nature. Chemical and electrochemical interactions with the media that come into contact with the surfaces can form an adhered layer. The properties of this layer may significantly differ from those of the bulk material. Friction on the surface can detach these layers, but they may also provide wear resistance, especially in metallic contacts. Once the oxide layer on metals is removed, it may not have enough time to fully form before the next cycle of a slider, so wear rises. Corrosive media can also penetrate cracks and pores, affecting the subsurface structure.

1.3.5. Fretting wear

In fretting, the oscillating relative motion between the two bodies is limited to an extremely small amplitude. In the case of fretting, wear debris can be trapped between the two surfaces, so the properties and the chemical composition of the materials between the two surfaces can affect the behavior of the tribo-system.

In most applications, these wear models do not apply due to the dynamic behavior in wear. This means, that the coefficient of wear as a function of time $K(t)$ is usually not constant.

1.4. Third Body and Running-In

In most applications, engineers need to deal with complicated systems with a strongly dynamic character. As mentioned in the section of fatigue wear, repeated loading of a surface may give rise to structures that differ from what would be expected after a single pass. The formation of wear particles, change of topography, material transfer, mechanical mixing, adsorption of molecules from the environment, chemical reactions and other parameters can significantly change the tribological behavior of a system in time. To date, no theory can combine all these parameters to describe the behavior and evolution of a tribo-system.

These changes are not always to be seen as negative effects. Indeed, the corrosive wear of metals is a destructive evolution of a surface and perhaps one of the first examples that come to mind. Still changes near the interface often lead to significantly improved tribological performance. In engineering the term "running-in" is used to describe the conditioning of sliding or rolling surfaces once contact is established. The duration of this process is different depending on the system and the conditions of load.

Thanks to advances in tribological testing methods, the evolution of the friction coefficient and the wear rate can be monitored during a sliding experiment. An example of the behavior of a sliding pair of metallic surfaces during running-in and when steady state is reached is shown in Figure 1.1. The friction coefficient and wear rate in the very beginning of a multi-pass sliding experiment is in some cases relatively low because of oxide layers that provide wear resistance. Once such a layer is removed, an increase

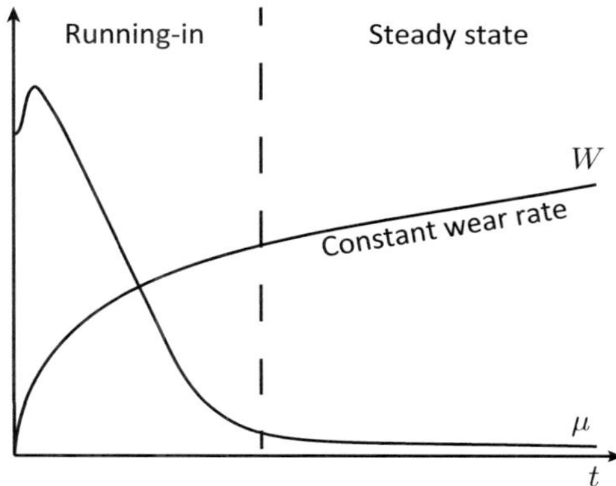

Bild 1.1.: Evolution of friction coefficient and wear during running-in and in the steady state.

of the friction coefficient and wear particle generation is observed. The running-in process has begun. Possible entrapment of particles between the surfaces may further increase the wear rate and friction force. Both the surface topography, as well as the near surface region undergo changes which lead to a system that features lower friction resistance and steady, low wear rate. It is usually accepted, that surfaces become smoother by the time the steady state has been reached. This is a general description of the processes that take place in a large number of tribo-system, however it is not observed identically in all cases.

It is often said, that during running-in, the "third body" is formed. Actually a wide variety of terms have been used for the formations observed, but the term "third body" will be used in this work.

So what is a third body?

In the 1970s, Godet made an effort to link lubrication theory and dry contact studies in one approach and cover aspects that interest both material scientists and mechanical engineers [19]. The two bodies that are in relative tangential motion are named the first bodies and anything separating the two, including the parts of the bodies near the interface that exhibit different properties from the bulk of the materials and the lubricant film constitute the third body. The third body transmits the load, accommodates the speed difference between the first bodies, and separates them, reducing their degradation.

What is usually referred to as third body in the case of lubricated sliding metal surfaces are the sub-surface structures that consist of the metal, its oxides, as well as a mixture of lubricant, its additives, and chemical products that form during sliding. The grains of the metal in this region are significantly smaller than in the bulk.

The running-in process and the formation of the third body are often too complicated to analyze. Numerous phenomena at different scales make the evolution of the near surface structure impossible to predict. This has introduced a great deal of empiricism in the study of tribo-systems.

1.4.1. Energetic approach

An interesting approach toward the better understanding of tribo-couples is an energetic approach. The energy dissipation expressed by the resistance against the relative tangential motion of two bodies that can be measured as the friction force. The work of this force is converted to several sorts of energy:

- Energy needed for **deformation** to occur. This can lead to plastic deformation, thereby moving dislocations, initiating and propagating cracks and changing the grain structure. The deformation can also be elastic and release the energy in other forms while relaxing. Wear is one of the most evident results of this energy channel.

- **Thermal** energy released in the system. At the tips of the asperities, the temperature may increase dramatically during contact, reaching the so-called flash temperatures.

- **Chemical** energy, which may often follow the release of thermal energy. This accounts for the chemical variations that are found in the third body. Friction can even initiate exothermic reactions, which in turn may start chain reactions near the surface.

- **Electric** energy, causing surfaces to become charged. In many cases this may also lead to chemical reactions.

- Emission of **photons**, which is often the result of discharge after increasing the electric energy.

- Emission of **phonons** that are usually absorbed in the solid bodies.

Examining the energy dissipation during friction is a very useful tool in order to trace the mechanisms that are activated during running-in and lead to the resulting third body [20, 21, 22]. An important asset when evaluating a tribo-system in an energetic approach is the use of the friction power P_F,

$$P_F = \frac{d}{dt} \int F_F ds = \mu F_N \upsilon, \tag{1.17}$$

with F_F being the friction force, μ the friction coefficient, F_N the normal force, and υ the sliding speed. If the energy released in a tribo-system is dissipated in a real volume V_r which is only a fraction of the system's total volume (first, third bodies, and the rest of the lubricant and wear particles), then the power density of friction ρ_F is the power per volume P_F/V_r. For simplicity, the energy dissipation channels mentioned above can be categorized in heat generation P_q, wear particle generation P_w, and others gathered together as P_{tb}, so the power balance should be

$$P_F = P_q + P_w + P_{tb}. \tag{1.18}$$

Despite the advantages of this approach, an energetic study alone fails to fully elucidate the dynamic behavior of a tribo-system. To achieve this, one has to break such a system down to pieces and study every process separately, understanding that it is only a part of a complex system.

1.4.2. Complexity theory

Complex systems abide by the laws of complexity, a field that emerged from the chaos theory in physics. They are usually very hard to understand and characterize completely. Some of the main characteristics of a complex system are:

- Non-linearity

- Emergence

- Interaction between properties

- Open systems

- Self-organization

- Attractors

Tribo-systems feature most of the above. The structures found in third bodies are clear marks of complex processes. Investigations of the topography and the subsurface structure show chemical variations, as well as formations like lamellar structures and vortical mixtures that indicate action of attractors. Furthermore, it is common knowledge that good running-in will form a third body with desired properties. This is proof of the ability of the system to organize itself toward the reduction of friction and wear. The final state is characteristic of the conditioning and cannot be altered into a different state unless severe wear removes the existing third body entirely. Apart from energy transfer according to the aforementioned mechanisms, material transfer is also to be considered. It can occur between first bodies, within a first body, into the lubricant as particles or molecules, and from the lubricant, including particles, molecules, and chemical byproducts.

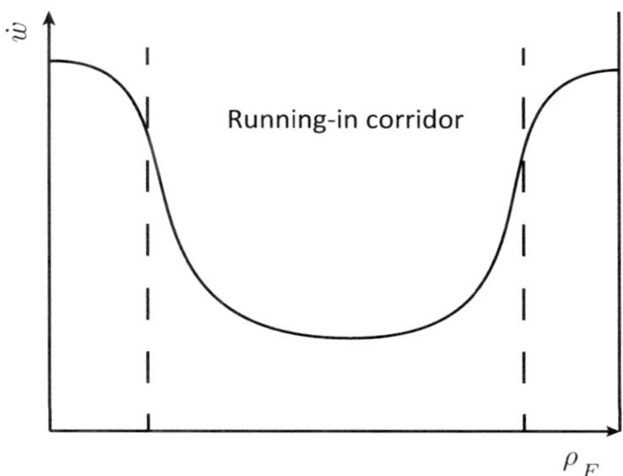

Bild 1.2.: Running-in corridor formed for low wear rates at the steady state depending on the power density of friction.

The complexity of third bodies and the dramatic changes that running-in can infer stress the importance of simplification. A solution is to study partial phenomena occurring during this process and trying to correlate their interaction with other parts of the system and the environment.

Running-in is not always as efficient. In lubricated sliding surfaces, it occurs in the boundary lubrication and mixed lubrication regimes, where plastic deformation is possible. Depending on the power density of friction, the final wear rate at the steady state is minimized for values that are not too high but also not too low. Too low ρ_F values lead to severe plastic deformation and high ρ_F does not allow the system to reach the optimal steady state. As shown in Figure 1.2 [23], a running-in corridor between extreme values lead to optimal final wear rates ($\dot{w}$), the time derivative of wear, which can refer to volume, mass or depth wear.

The properties of the third body and phenomena that relate to its formation are further discussed in Chapter 3.

1.5. Lubrication

The use of media to separate the two surfaces in relative tangential motion in order to reduce friction and wear is very common and often crucial for a large number of applications. These media can be fluids or solids. Despite the importance of solid lubrication in modern engineering, the main focus of this work is placed on liquid lubrication. The greatest part of this section is devoted to the principles of liquid lubrication and it's regimes. The properties and role of solid lubrication are gathered in brief in the following section.

1.5.1. Properties of liquid lubricants

Liquid lubrication is usually performed using mineral and synthetic oils. Petroleum refining processes provide a wide variety of lubricant sorts [24]. These products are often further modified with additives in order to achieve the desired characteristics for every application. Generally, the properties of a lubricant can be separated in the properties of the base oil and those of the additives. In this work, only base poly alpha olefin (PAO 8 - the number eight refers to the cSt kinematic viscosity of the oil at $100\,^{\circ}C$) was used as lubricant, so additivation will not be examined in further detail at this point.

Polyolefins are rather simple synthetic lubricants, which are produced by alkenes (C_nH_{2n}). An example is shown in Figure 1.3. Poly alpha olefin in particular, is made by polymerizing an alpha olefin, an alkene in which the carbon-carbon double bond is located at the α-carbon atom. The plethora of alkyl branched groups of the polymerized molecules lead to various conformations, which prevents their crystallization even at low temperatures, as they cannot align next to each other easily. This way they remain in an oily and viscous state. An interesting property of these oils is that they are transparent and colorless.

A lubricant, be it an engine oil or grease for large scale excavators, plays an important role in the physics and chemistry of tribo-systems. Molecules

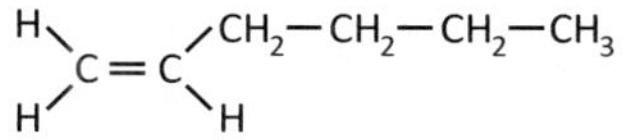

$$H_2C=CH-CH_2-CH_2-CH_2-CH_3$$

Bild 1.3.: 1-hexene, an example of an alpha olefin.

adsorbing on surfaces may seriously affect the nature of a contact, perhaps even change the chemistry of the solid body in the vicinity. The viscosity of a lubricant is one of the fundamental features that need to be examined when applying a lubricant. It is also important to consider the temperature dependence of viscosity. In addition, lubricants need to cover the surfaces of the sliding bodies, so the surface energy can also play a key role in lubrication. Finally, depending on the application, an engineer should consider some additional factors, such as density, fire resistance, pour point and a variety of chemical properties. As mentioned above, many of these characteristics in a lubricant are tweaked with additives to achieve the desired features.

1.5.2. Regimes of liquid lubrication

One of the main aims of the research conducted in this work is to study the dynamics of sliding systems in lubrication. A main difference between dry and lubricated friction is the formation of a film that keeps the two surfaces apart, reducing the real area of contact, or even completely separating the two surfaces from each other. A system may therefore respond very differently mainly depending on the thickness of such a film.

The friction coefficient in this case consists of two components. One that corresponds to friction generated at the real contact area and another from the flow of the film between the surfaces. The shear stress τ_{Fl} exerted by the fluid according to the Newtonian behavior is calculated as follows,

$$\tau_{Fl} = \eta \frac{dv}{dh}, \tag{1.19}$$

where η is the viscosity of the fluid and dv/dh is the velocity gradient perpendicular to the direction of shear. This is a simplified case, where turbulent flow between rough surfaces is not considered. Furthermore, oils often exhibit non-Newtonian properties, which makes the study of their behavior complicated.

Depending on the contribution of each term, four regimes can be defined:

- **Boundary lubrication (BL):** Collisions between asperities of the two surfaces produce friction, heat and wear. The friction coefficient depends mainly on the mechanical properties and nature of the junctions.

- **Mixed lubrication (ML):** Some asperities deform plastically and some only elastically when they come in contact with asperities of the counter-face.

- **Elastohydrodynamic lubrication (EHD):** Many asperities still deform elastically when they come in contact with asperities of the counter-face. The properties of the fluid affect the friction coefficient along with the mechanical properties of the contacts.

- **Hydrodynamic lubrication (HD):** The surfaces are practically apart, so the friction coefficient depends mainly on the film separating them.

A common representation of these regimes is that of a Stribeck curve [25], a diagram that shows the friction coefficient in steady state depending on a parameter calculated by the viscosity multiplied by the relative speed over the normal load (p_N). Such a curve is shown in Figure 1.4.

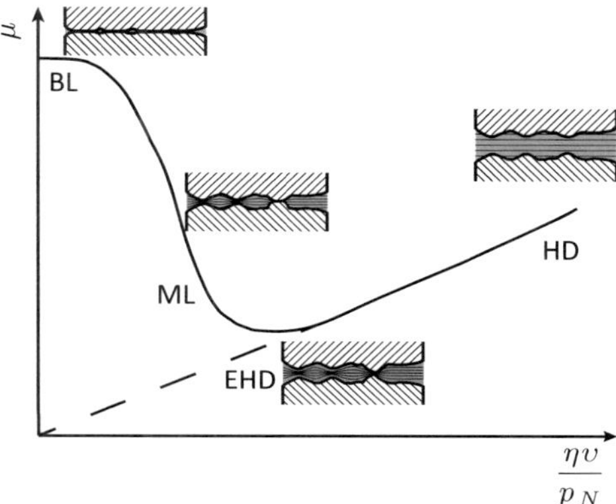

Bild 1.4.: Stribeck curve showing the four lubrication regimes.

1.6. Modern Experimental Methods for the Examination of Tribo-systems

Tribology is a particularly diverse science and a very interdisciplinary field of research. This diversity is also reflected in the wide variety of instruments that are used for tribological investigations. This section contains a brief overview of methods that are important for the analysis of both the surface and the sub-surface of sliding systems, as this is the main subject of the present work.

1.6.1. Tribometers

Tribometers may vary in form, but among the measured values, normal and friction forces are always monitored during the experiments. It is hard to categorize these instruments.

They can be classified according to the size of the contact:

- **Nanotribometers:** The most common tool for investigations at the nanoscale are friction force microscopes. Atomic force microscopes in different configurations can be used depending on the variables that need to be examined. MEMS can also be used for direct tribological measurements, usually aiming at simulating real application conditions.

- **Microtrobometers:** Sliding reciprocating micro-tribometers usually use a sphere against a flat surface. Nevertheless, they could also be optimized for measurements of small contact areas between less simple geometries such as sphere against spheres, or even the tip of a toothbrush fiber against a tooth [26].

- **Macrotribometers:** There are many instruments that fall into this category. Pin-on-disk tribometers are probably the most common. Roll on plate, prism tribometers and a wide variety of machine simulators can also be found. Measuring gauges can be integrated in machine parts allowing for direct measurment in an application.

- **Large scale machines and simulators:** This category includes machines with incorporated sensors as in the above cases, but measurements take place at significantly larger scales.

The environment of the experiment may also differ among devices. Measurements are conducted at:

- **Ambient conditions:** These are the most common and usually the easiest to operate tribometers. In some cases, some conditions have to be controlled, such as the temperature.

- **Controlled humidity:** Controlling the humidity of the chamber where the experiment takes place may significantly change the behavior of the tribo-system, as water from the atmosphere may act as a lubricant, or create a meniscus changing the apparent normal load. Both adsorbed and absorbed humidity, for example in a porous material, can affect the friction coefficient.

- **Controlled gas environment:** Introducing gases may affect the chemistry of the surface, possibly leading to reactions that affect the third body. Inert gases can prevent the formation of oxides on metallic surfaces.

- **Vacuum:** As mentioned above, gases and humidity may alter the behavior of a tribo-system. In many cases, the contacts of pure surfaces need to be examined, so vacuum or perhaps even ultra high vacuum is necessary. Recently tribology made it to outer space, as tribometers were even installed on satellites to determine the properties of tribo-systems in vacuum, zero gravity and direct exposure to the sun.

- **Lubrication:** The importance of lubrication has been stressed extensively above. Different tribometers with flowing, stagnant lubricant, or even a thin lubricant film can be used to better understand the regimes of lubrication and their effect on different tribo-systems.

Finally tribometers can be categorized according to measurements running-parallel to the main recording of the forces. Such examples can be temperature measurements, contact area with optical methods through transparent counter-faces, electric current at contacts, chemical changes in the lubricant, etc. An important breakthrough is the simultaneous measurement of wear via radionuclide technique (RNT) [27] in liquid lubrication.

1.6.2. Other wear measurement techniques

When wear occurs, wear particles are released and carried away with the flow of the lubricant. If the surface that undergoes wear is marked with radioactive nuclides, then the concentration of radioactive material in the lubricant increases. If the volume of the fluid and the area of sliding contact are known, then the wear rate can be calculated on-line, usually in µg/h which can be translated into nm/h. More than one active species can be measured at the same time, allowing for simultaneous on-line measurement of wear on both surfaces.

An RNT unit consists of a NaI detector in a protective Pb chamber and a pump that conveys the lubricant with the wear particles from the sliding contact to the detector and establishes lubricant circulation between the two. An additional detector can be used to measure a reference sample.

At least one of the two surfaces needs to be labeled before performing such an experiment. This is one of the most crucial steps in the implementation of an RNT experiment. Small parts can be activated by placing them directly in the neutron flux of a nuclear reactor, leading to a homogeneous distribution of radioactive nuclides throughout the part volume. Cyclotron accelerators can direct protons, deuterons, or α-particles onto a predefined surface with a mask. Depending on the penetration depth, the collision of these particles with the atoms of the sample yield a thin activated area, the thickness of which can reach several hundreds of nm. This results in a significantly lower activation which is limited in a desired area, thereby making the sample easier to handle in terms of security measures. For high wear rates, the gradient of radionuclides in depth beneath the surface has to be considered for the calculation of wear rates. The precision it offers is unique for the simultaneous monitoring of applied forces and wear.

1.6.3. Topography measurement

Different techniques can be used to gather data on the topography of a surface:

Atomic force microscopy: A small probe with an atomic scale tip is supported on a cantilever that scans the surface of a sample, while a reflected laser beam at the rear of the tip reveals the vertical displacement. Atomic force microscopes can be used in both contact and non-contact mode to acquire 3-D images of a surface. A variety of configurations are available to target specific surface forces (atomic, Van der Waals etc.), which can be

Tabelle 1.1.: Comparison of topography measurement methods commonly used in tribology (representative values gathered from instruction manuals, instrument software and instrument suppliers' web pages).

	AFM	Tactile Profilometry	Laser Profilometry	Confocal Microscopy	Interfero metry	Holography
Vertical Resolution	0.01 nm - 20 µm	10 nm - 1 mm	10 nm - 10 mm	1 nm - 100 µm	0.1 nm - 50 µm	0.1 nm - 50 µm
Lateral Resolution	0.1 nm	1 - 30 µm	1 µm	0.5 µm	0.5 µm	0.5 µm
FOV	100 nm - 80 µm	1 - 50 mm	1 - 5 mm	20 µm - 1 mm	20 µm - 1 mm	20 µm - 1 mm
Measurement Time	min - h	sec	min - h	sec - min	sec - min	msec

(FOV is the field of view)

attractive or repulsive. The resolution of these instruments can even reach atomic levels under certain conditions.

Tactile profilometry: Similarly, a significantly larger probe, in the range of µm to mm, or a stylus, comes to contact with a surface and a line is scanned. The vertical displacement of the probe can be measured in various ways.

Laser scanning profilometry: A laser beam is directed on the surface and it scans lines with the assistance of a planar positioning stage.

Confocal microscopy: A laser beam is focused on a sample and the reflected light is directed via a beam splitter and mirrors to a pinhole, that only allows in focus light to reach the detector. Thanks to a motorized motion of the sample, or the mirrors, in focus information is gathered for every plane of the sample leading to the construction of a 3-D image.

Interferometry: Again, a motorized motion allows for a full height scan of the sample, only here, the constructive and destructive interference of the initial cohesive light beam and the reflected beam is used instead of a pinhole. To avoid mistakes in reconstruction from height leaps greater than the wavelength of the light source, a distribution of wavelengths composing white light can be used.

Holographic microscopy: The coherent monochromatic light is again split, part of it is reflected at the sample. A beam splitter rejoins the reflected beam with the initial beam and directs them to the detector. A motorized motion is not necessary, as the reconstruction of the 3-D image is performed with a numerical reconstruction algorithm.

1.6.4. Electron microscopy and focused ion beam

In electron microscopy, the light is replaced by a monochromatic electron beam emitted by a tungsten filament or a field emission gun. These are focused with electromagnetic lenses onto or through a sample. By moving the beam around, an area of the sample can be scanned.

In scanning electron microscopy (SEM), electrons interact with the electrons of the atoms near the surface of a sample. The electrons of the primary beam may then be backscattered (elastic scattering), completely lose their energy in the sample, or lose part of their energy and find their way out. Back scattered detectors give material contrast information on the surface, as the intensity of the beam depends on the atomic number of the atoms. The electrons that have lost part of their energy are accelerated toward a positivley biased grid and then to the secondary electron detector. The second contrast mechanism gives the impression of a topography image, without, however, measuring height information.

Tabelle 1.2.: Comparison of surface chemical analysis methods commonly used in tribology (representative values gathered from instruction manuals, instrument software and instrument suppliers' web pages).

	EDX	**AES**	**XPS**	**FT-IR**
Detected Species	Elements Z>4	Elements Li - U	Elements Z>2, Chemical structure and bonding	Chemical Structure and bonding
Penetration Depth	2 μm	1 nm	1 - 10 nm	5 - 10 μm
Mapping Lateral Resolution	2 μm	10 nm	5-8 μm	-

In transmission electron microscopy (TEM), the thickness of the sample has to be smaller than the penetration depth of the electrons. Samples can be prepared by mechanical milling, chemical etching, or ion etching. Depending on the position of the sample and the detector with respect to the focal and imaging planes, different contrast mechanisms are used, giving information on the thickness of the sample, diffraction patterns, or imaging of the crystallographic sample.

A focused ion beam (FIB) can be used to mill a surface. Usually Ga ions are directed onto a surface in the same way as electrons. As ions are significantly larger than electrons, their interaction with the surface is destructive. In this process, surface atoms are removed, so a pit, or section can be created. During collision, ions, neutral atoms and electrons are released. Ions and electrons can be used to generate contrast in a depth profile. Alternatively, the side-wall of a pit formed during etching can be examined with electrons by tilting the sample. In channeling mode, electrons can provide contrast between grains, depending on their orientation.

1.6.5. Surface chemical analysis

The topography and grain structure of a tribo-system are very important pieces of the puzzle, but the image is still not complete. The chemical composition can be just as important in understanding the third body. As the volume that is usually affected is near the surface, the following methods are among the most important chemical analysis tools for tribologists:

- **Energy dispersive X-ray spectroscopy (EDX):** This method is usually utilized in conjunction with SEM. When a focused electron beam interacts with a specimen, some of the electrons orbiting the atoms are excited or removed from the inner shells. Other electrons will change state to occupy the inner shells, emitting X-ray light at wavelengths that are characteristic of the element.

- **Auger electron spectroscopy (AES):** This is a variant of the above method, in which the energy released during the transition of an electron to a hole in the inner shells is absorbed by an electron of the outer shells. Its kinetic energy is again characteristic for the mechanisms that occur in different elements. These two methods have the advantage of using the electrons of the SEM as probes, so the affected spot on the surface is very small. Therefore these methods lend themselves to high resolution surface element mapping.

- **X-ray photoelectron spectroscopy (XPS):** In this case, the initial excitation of electrons is induced by directing an X-ray beam onto the specimen. The excited electrons of the atoms are completely removed and their kinetic energy can be used to determine the element of their origin. Although the affected spot is larger than in the methods mentioned above, this method has the advantage that all elements with more than two electrons (He) can be detected. Furthermore, the sensitivity of this method can be utilized do identify peak shifts due to chemical bonds. Finally this method is often combined with ion beam etching to examine the subsurface composition.

- **Fourier transform infrared spectroscopy (FT-IR):** The methods mentioned above are very efficient in isolating near surface chemical information; however, they can only be used when high or ultra high vacuum is available. If this is not an option, samples can alternatively be examined with traditional spectroscopy. Infrared spectroscopy is a very common tool for chemical analysis and some of its variants, such as the attenuated total reflectance (ATR) FT-IR can also be very efficient in isolating surface information, provided that the surface is relatively homogeneous.

- **Other useful analytical methods:** Further methods often used in tribology may include destructive methods or variants of some of the methods mentioned above. An example is Glow Discharge Optical Emission Spectroscopy (GDOES), which involves sputtering of a flat surface to quantify the composition of metals, metallic alloys, or intermetallic phases on a surface or a coating. Raman spectroscopy relies on inelastic scattering and is often employed to specify chemical bonds and symmetry of molecules, as well as the crystallographic orientation of a grain. In some cases it can be used to detect stresses. Finally, a method that can be used to detect metals is inductively coupled plasma atomic emission spectroscopy (ICP-AES).

In-situ approaches combine more of the above techniques simultaneously. References of such applications are given in chapter 2.

2. Design and Construction of the Tribometer

2.1. Introduction

For the better understanding of fundamental processes in tribology it is important to characterize the interface, which is formed by the two sliding bodies and a possible lubricant. Because this is experimentally very challenging, tribologists often have to rely on simulations. Remarkable advances in this field have been made in the last decades [28, 29]. From simple contact cases to lubricated surfaces, simulations cover a wide range of conditions. Despite these advances, scientists are still in need of scientific experimental results in order to continue delving deeper into the phenomena observed.

A pin-on-disk tribometer and a topographical microscope can be considered common assets of a tribological experiment. Nevertheless, in the vast majority of such experimental procedures, the samples are subjected to several cycles of frictional loads prior to being examined with a microscope. Therefore, they provide no information about the dynamic processes that occur during these cycles. Normal and lateral forces can be recorded during the experiment, but topography and wear are only observed afterwards.

An interesting approach toward studying sliding surfaces *in-situ* is observing one of the two surfaces through the body it is sliding against. This requires that at least one of the two bodies is transparent [30, 31]. The main advantage of this method is that the images acquired, are images of the actual contact area. On the other hand, transparent materials are rarely encountered in demanding tribological systems, so this method poses great limitations on the variety of samples that can be used.

Recently, an effort was made to monitor the wear scar topography after each load cycle. The instrument utilized for these experiments comprises a linear positioner, a white-light interferometer (WLI), and a pin [32]. This instrument can even conduct experiments in different gas atmospheres in a closed chamber; however, it is limited to a linear reciprocating mode and cannot monitor surface changes under lubrication.

While the measurement of *in-situ* topography in a tribological experiment is still being developed, wear particles can be measured using commercial systems. The highest resolution is currently achieved with (RNT) instruments (see Chapter 1).

2.2. Experimental

2.2.1. Principle of function and instrumentation

In this work, a new concept for the on-line monitoring of the topography and wear measurement of lubricated systems is introduced. This is achieved by means of a novel tribometer that consists of state-of-the-art instruments for planar positioning, surface topography, wear measurement and force sensing.

A high precision planar positioning system provides translational freedom on the xy plane. A PPS 200-4 system of Tetra GmbH, Germany, was employed for this purpose. The device has a range of $200\,\text{mm} \times 200\,\text{mm}$ and reaches a maximum speed of $500\,\text{mm/s}$ with ± 1 μm precision. The position is monitored at a 20 nm resolution.

A tank which holds a flat plate sample is placed on top of the positioner which drives it along the desired path. The tank is connected to an RNT (Zyklotron RTM 2000 of Zyklotron AG, Germany) apparatus performing on-line wear measurements. Three measuring devices are attached in three fixed positions over the sample: a force sensor, a holographic microscope and an atomic force microscope (AFM).

Digital holographic microscopy is an interferometric technique which can be used to focus numerically on different object planes without any necessary movement of optical parts. This makes its use in immersion mode easier, especially when there only is a thin film of liquid on the sample. Furthermore, the acquisition of an image is significantly faster and the system lends itself for automated processes. The holographic microscope used in this case is a DHM® R1000 series (Lyncée Tec SA, Switzerland) in a customized case that fits the present instrumental layout. The objective lenses employed are a $50\times$ oil immersion and a $10\times$ dry. This device can acquire images at a maximum frequency of 15 frames/s. The software of this instrument was adjusted for the optical properties of the lubricant, in this case transparent poly alpha olefin (Fuchs Petrolub AG, Germany). The AFM utilized here is an ULTRA Objective (Surface Imaging Systems / now Bruker AXS Microanalysis GmbH, Germany) with a scanning range of $80\times80\,\mu m^2$ and an additional kit for immersion measurements. In order to measure forces, a customized setup was built, which includes three SKL1417-IR (Tetra GmbH, Germany) fiber optic sensors (FOS).

As the sample moves, the sensors are focused at different positions on the sample surface. The relative position of all instruments is designed in such a way that all devices have access to a specific part of the path. The schematics of this concept are shown in Figure 2.1. A pin is attached at the bottom of the force sensor. With a closed path without angular geometries, the pin can perform smooth cycles without losing contact with the plate-sample. Part of the wear track left behind can be examined with the other two sensors by pausing the motion very shortly. This allows for acquisition of surface topography images after each cycle at the exact same position every time.

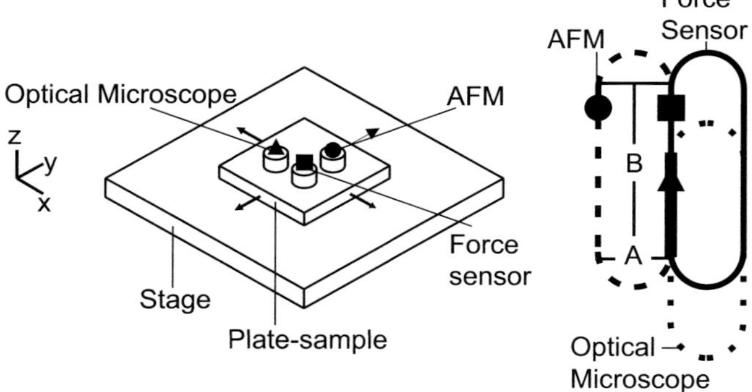

Bild 2.1.: According to the concept of the present instrument, the sample has translational freedom on the xy plane, limited by the size of the stage. The force sensor with the pin (square), the optical / holographic microscope (triangle) and the AFM (circle) are attached at fixed positions over the sample. For a race-track shaped motion: the dashed, dotted and thin continuous lines on the right show the paths on the sample accessible to the AFM, optical microscope and force sensor respectively. The thick continuous line is accessible to all sensors.

2.2.2. Mechanical setup

The base of the tribometer is a granite plate that stabilizes the structure and minimizes height variations. The positioning system is placed directly on the plate and holds the oil-tank. The plate sample is located inside the oil-tank with a polished surface facing up. Two bridges, one made of granite and another one of aluminum profiles, hold the rest of the sensors above the sample. The granite bridge is necessary in order to reduce vibrations that affect the AFM and force sensor measurements. On the other hand, the holographic microscope is less sensitive to vibrations, so a plain aluminum construction is sufficient. The layout of the AFM, force sensor and holographic microscope over the sample matches the one described in the instrumental concept (Figure 2.1).

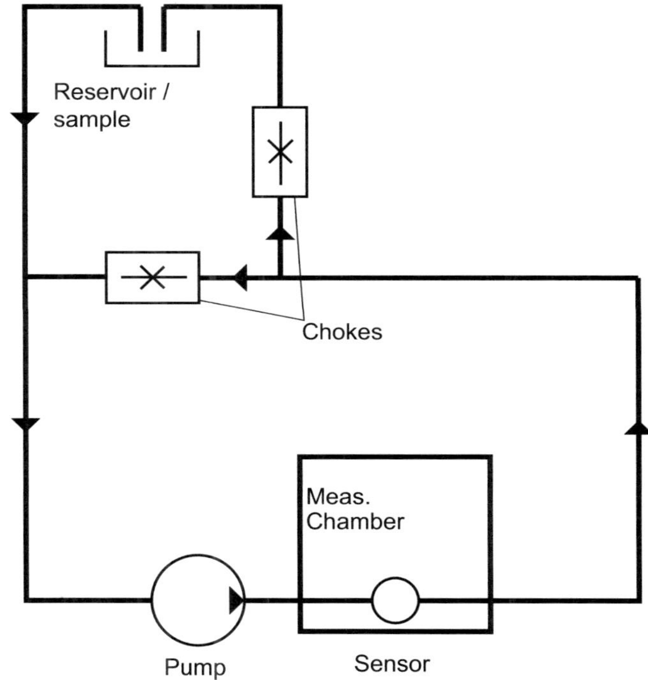

Bild 2.2.: Oil circulation diagram. The pump is followed by a radioactivity sensor. Both are located in the RNT instrument. Outside this apparatus, the flow is split to access the oil container / reservoir and forms a bypass circuit. Two chokes control the flow in each direction. The sample is located in the reservoir area.

2.2.3. Oil circuit

The oil circuit connects the tribometer with the on-line wear measurement system. A diagram of the oil circulation is shown in Figure 2.2. The oil-pump and the radioactivity sensor are located in the RNT apparatus. A bypass circuit is controlled by two chokes, limiting the oil flow at the inlet and outlet of the oil container/reservoir. Depending on the volume of the liquid in the reservoir, the total oil volume varies between 1 and 2 L.

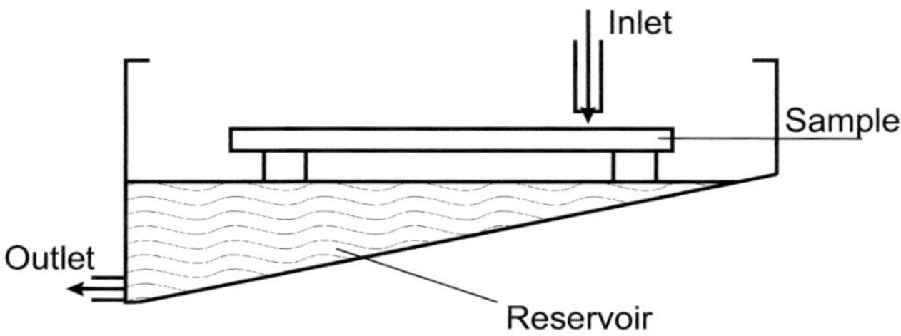

Bild 2.3.: The sample is located above the oil reservoir. The lubricant flows onto the sample and is collected from the lowest point of the reservoir.

The sample is placed in an oil container; however, it remains above the oil level, as shown in Figure 2.3. The lubricant flows onto the sample and carries the particles off the surface. Continuous flow replenishes the reservoir. The bottom of the container is inclined in order to reduce the volume of oil in the reservoir. This improves the resolution of the wear measurement and reduces the inertia of the positioning system. The total mass of the container and the sample without oil is between 7 and 10 kg. The positioning system sets a limit of 150 N normal force. Finally, a tube at the lowest point of the container leads the oil back to circulation.

2.2.4. Force sensor

The force sensor block consists of an up-scaled version of the so-called "tribolever" [33] design, three FOS and a steel scaffold holding them together. Because of its small size, the original tribolever comprises a pyramid on top to support the displacement measurement mirrors. In the design presented in this work, the pyramid is replaced by a cube. Three adjacent faces of this cube are occupied by mirrors: two sideways (x, y) and one on top (z). The FOS are facing against the mirrors and measure displacements of the cube in the x, y and z direction. Finally, a pin is screwed at the bottom of the cube. Sketches of the force sensor block can be found in Figure 2.4.

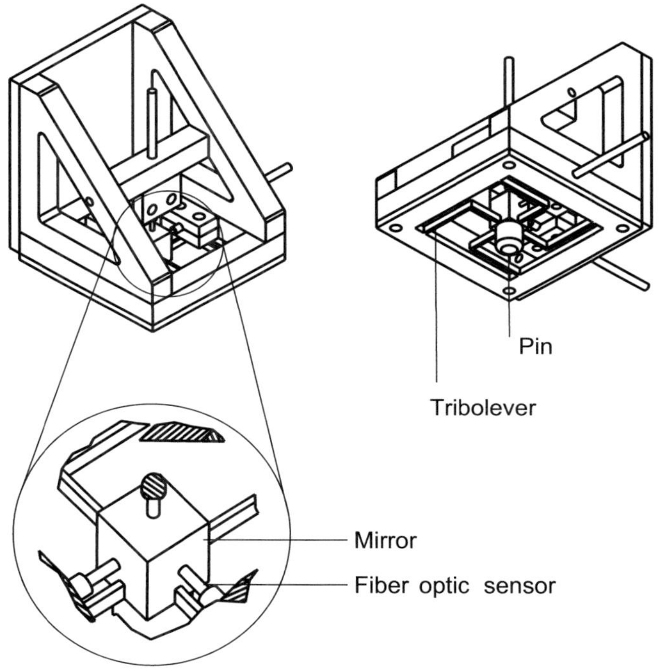

Bild 2.4.: A customized scaffold holds the force sensor. The pin is screwed at the bottom face of the central cube, while three mirrors are attached on the top and two adjacent side faces. Fiber optic sensors are fixed against them and measure the x, y and z displacement.

Two important factors that had to be considered for the up-scaling were the applied forces and the resulting displacements. The aim was to optimize the sensor for conventional tribological systems, in which case the frictional force is only a fraction of the normal force. Therefore, the ratio of the spring constant in the z direction (k_z) over either one of the lateral directions (k_x, k_y) should be of the same order of magnitude as the friction coefficient. The applied normal force should result in a pressure that allows for running-in. The desired contact area for the present instrument is of the order of tens

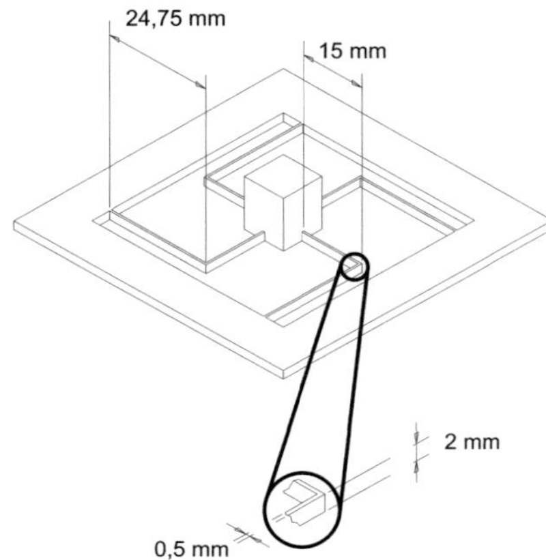

Bild 2.5.: The force sensor is an up-scaled version of the "tribolever". The dimension and aspect ratio of the legs have been optimized for macroscopic tribological experiments..

of mm^2. Finally, the displacements should be suitable for the detectors (the FOS employed for this apparatus have a range of about 800 m).

As mentioned above, the up-scaled version of the so-called "tribolever" was designed for the measurement of the normal and both lateral forces. The main reason for selecting this geometry is that it provides equal sensitivity for both lateral directions while remaining relatively insensitive to torsional forces. The selected material was steel, with a Young's modulus of 210 MPa.

Although the proportions of the construction should remain close to the original design, some modifications were necessary. Most importantly, the applied forces and the resulting displacements were in a different order of magnitude, the size and mass of the block supporting the mirrors differed from the original tribolever and the material used was not Si. To determine the optimal geometry for the up-scaled version of the sensor, the part was simulated with a Finite Element Model (Simulia$^®$ Abaqus).

Bild 2.6.: Meshed part as imported from CAD design for simulation with Abaqus.

In this simulation, steel with a Young's modulus of 210 GPa and Poisson's ratio of 0.3 were set. The original design of the up-scaled tribolever is shown in Figure 2.5. In order to simplify meshing, the pin was replaced by a cube identical to the one that supports the mirrors. The mesh was based on 0.5 mm cubes. Figure 2.6 shows the part as imported in the simulation with its mesh structure. The square frame that supports the legs was defined as a constant fixed structure. All forces were applied in the center of the square at the base of the lower cube. The position of five reference points was monitored. These points were placed in the center of every free face of the top cube. They include the positions where the mirrors are fixed, in order to measure displacements.

To examine the sensor's response to forces in different directions, four sets of simulations were performed:

1. Set 1: $F_x \neq 0$, $F_y = F_z = 0$.

2. Set 2: $F_x = F_y$, $F_z = 0$.

3. Set 3: $F_x = 0,05F_z$, $F_y = 0$.

4. Set 4: $F_x = F_y$, $\left| \vec{F_x} + \vec{F_y} \right| = 0,05F_z$.

Tabelle 2.1.: Simulation Parameters

Set/Test	F_x [N]	F_y [N]	F_z [N]
1a	0.5	0	0
1b	1	0	0
1c	2	0	0
2a	0.345	0.345	0
2b	0.707	0.707	0
2c	1.414	1.414	0
3a	0.5	0	10.012
3b	1	0	20.025
3c	2	0	40.050
4a	0.567	0.567	16
4b	0.707	0.707	20
4c	1.414	4.414	40

The tests cover uni-axial load in the sliding direction (x), load in 45° to the cube's face (xy) and finally these two cases combined with a normal force (z). In sets 3 and 4 the friction forces were set to 0.05 times the normal force. This corresponds to a representative friction coefficient of lubricated metallic surfaces after running-in. Table 2.1 shows all force configurations for every test.

Figure 2.7 shows a graphical representation of the sensor's response to the application of forces as described in set 4. The deformation has been overscaled to demonstrate how the legs bend.

The response of the sensor to the application of forces was linear, as expected for the settings of the simulation. The x displacement was calculated as the difference between the final and the initial position of the center of the face of the cube perpendicular to the x-axis on top of the sensor. This is the point, where the FOS (Fiber Optical Sensor) is focused to measure distances. A linear fit was applied to the resulting data points of set 1 to determine the spring constant of the sensor. Equation 2.1 results from this fit. The calculated spring constant is 782.60×10^3 N/m. The results for sets

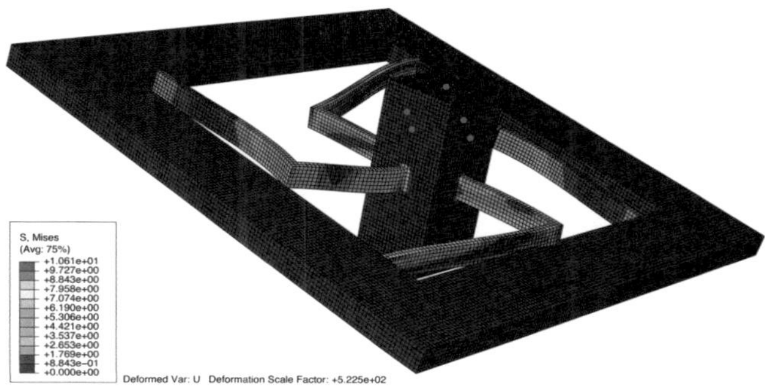

Bild 2.7.: Deformed part as as simulated with Abaqus with the force configuration described in set 4.

2, 3 and 4 are 787.77×10^3 N/m, 782.22×10^3 N/m and 781.53×10^3 N/m respectively.

$$F_x = 2.35 \times 10^{-6} N + 782.60 \times 10^3 N/m x \tag{2.1}$$

Where F_x is the applied force on the x-axis and x the displacement on the x-axis.

The displacement on the y-axis was also calculated for sets 1 and 3. In these two cases no force was directly applied on the y-axis, so this measurement only demonstrates how the application of a force along the x-axis causes a displacement along the y-axis. Similar linear fits were applied for these values, but in this case the displacement and the applied force were perpendicular to each other. The resulting constants on the y-axis were -10^9 N/m and 666785×10^3 N/m for sets 1 and 3 respectively. The absolute values are very high; therefore, they are negligible when measuring the displacements caused by forces directly applied on the y-axis.

Finally, the rotations around the x, y and z axes were calculated. A maximum angle of $0.025\,^\circ$ resulted on the y-axis, when 2.828 N, 2.828 N and 80 N were applied in the x, y and z axes respectively. Such small angles are

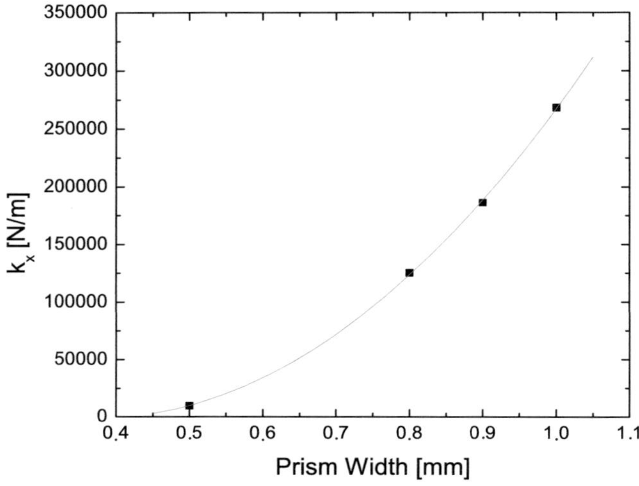

Bild 2.8.: Dependence of k_x on d as simulated with Abaqus. The result of the fit is given by the following equation: $k_x = 92742.0 - 505576d + 680345.9d^2$.

not expected to cause considerable deviations in the measurements with the FOS.

The tests described above were repeated for different leg geometries. The width (d) varied from 0.5 to 2.5 mm. The Hooke's constants (k) for the x, y and z axes were calculated for each case. The dependence of k_x and k_z on d is shown in Figures 2.8 and 2.9. Polynomial fits were performed to express the relation between these values.

In order to achieve high geometrical precision, the force sensor was machined by means of spark erosion cutting. A 10 mm wide border of the initial plate was preserved around the legs to support and stabilize the structure as shown in Figure 2.4.

After constructing the first sensor (d=1mm), a calibration was performed by hanging measured weights along the x, y and z axes at the position where the pin is attached. The displacement measured with the FOS for each load applied was used to calculate k_x and k_z. Figure 2.10 shows a comparison

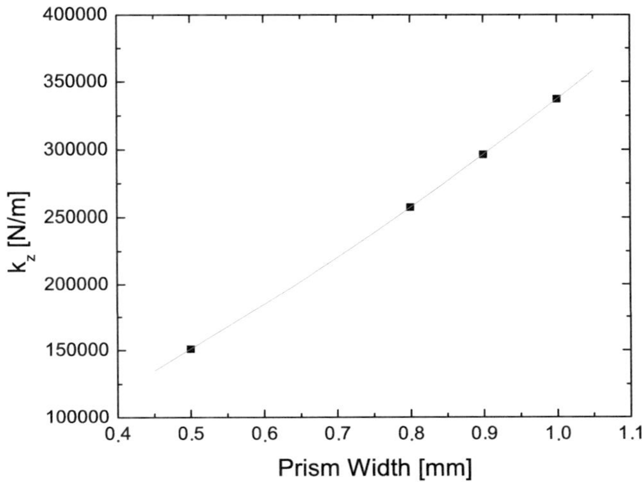

Bild 2.9.: Dependence of k_z on d as simulated with Abaqus. The result of the fit is given by the following equation: $k_z = 9278.0 - 238872.0d + 89159.5d^2$.

between the results of the calculation with the measured values. There is a remarkable difference between the slopes of the linear fit of the calibration and that of the simulated data. This result implies a significantly lower E Modulus for the machined part as compared to the simulated component. The reason for this deviation was not investigated in detail; however, the discrepancy cannot be merely attributed to material properties. An image of the sensor's surface and manufacturing precision is shown in Figure 2.8.

Considering the above, the simulated data could not be used for a precise decision with respect to the geometry intended for the applied normal and lateral forces. Nevertheless, the trends shown by the polynomial fit of Figures 2.8 and 2.9 were used as guidelines for the construction of further sensors to cover a wider force range.

To facilitate the aforementioned wider range of force measurements, two separate sensors were constructed. These have different width and thickness of the four legs (A and B as shown in Figure 2.5) which result in dif-

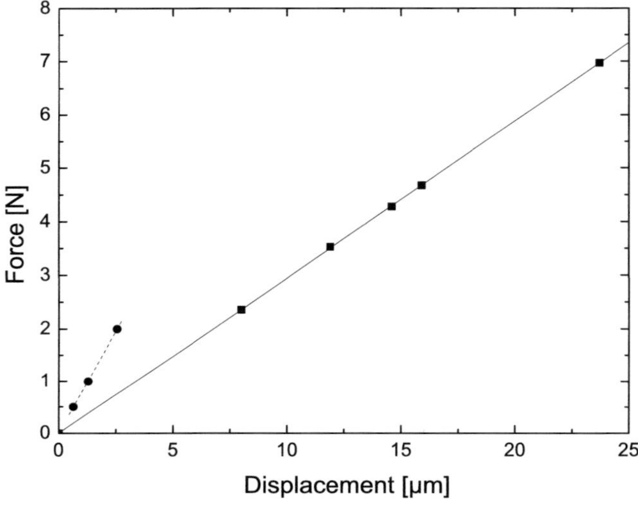

Bild 2.10.: Comparison between the values of kz from the calibration and the simulation. The circles represent the results of the simulation and the squares those of the calibration performed with measured weights.

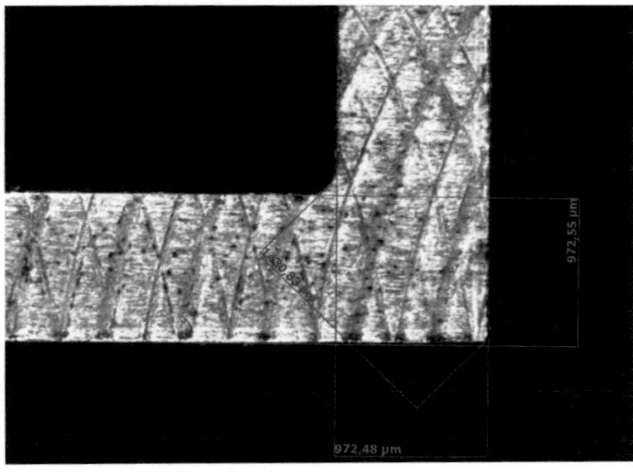

Bild 2.11.: Force sensor's machining details. This image shows the 90° angle at one of the legs.

Tabelle 2.2.: Dimensions (A: thickness and B: width as shown in Figure 2.5) and spring constants (F_n: normal force and F_f: frictional force) for the two force sensors S_1 and S_2.

Sensor	k_f [N/mm]	k_n [N/mm]	A [mm]	B [mm]
S_1	9.45	20.90	0.5	1.0
S_2	288.25	244.40	1.0	2.0

ferent spring constants. The dimensions and constants of the two sensors are summarized in Table 2.2.

2.2.5. Samples geometry

Two samples are required for a tribological experiment with this equipment. A tablet shaped sample that is fixed at the tip of the pin and a plate sample that is placed on the positioning stage.

The pin consists of a screw adjusted to the bottom of the force sensor and a spherical sample holder (Figure 2.12). The sample holder allows for rotation with a maximum angle of $9°$. In contact, this enables the tablet shaped sample to be aligned parallel to the tablet surface. the diameter of the sample is 2 mm and the height 1.5 mm. The edge around the circular contact area is tapered.

Two different types of flat plate samples can be used. A large 200×200 mm^2 plate that can be fixed directly in the oil tank with screws, or flat samples of other geometries, smaller in size, that are attached to an underlying 200×200 mm^2 plate with an adhesive substrate on it. In both cases, the samples move along with the oil tank and only have translational freedom, controlled by the positioning system.

2.2.6. Software and control of the instrument

In order to perform on-line measurements it is necessary to synchronize all devices. Figure 2.13 shows the schematic of the software control. Positioning and force monitoring are performed via a LabView (National In-

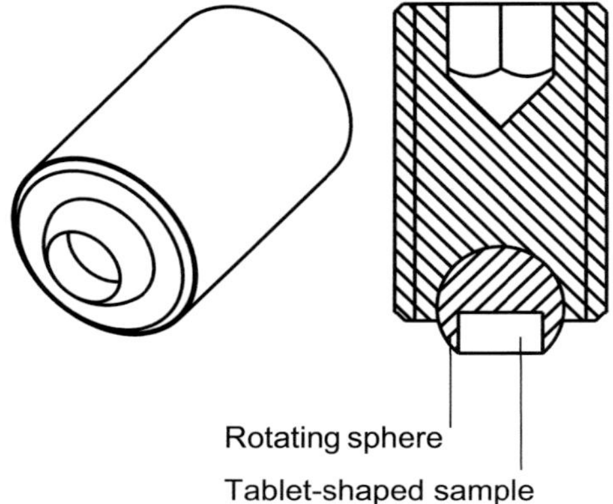

Bild 2.12.: The screw-pin with a rotating sphere. The tablet shaped samples are inserted in the sphere, which allows a max. of 9 ° rotation.

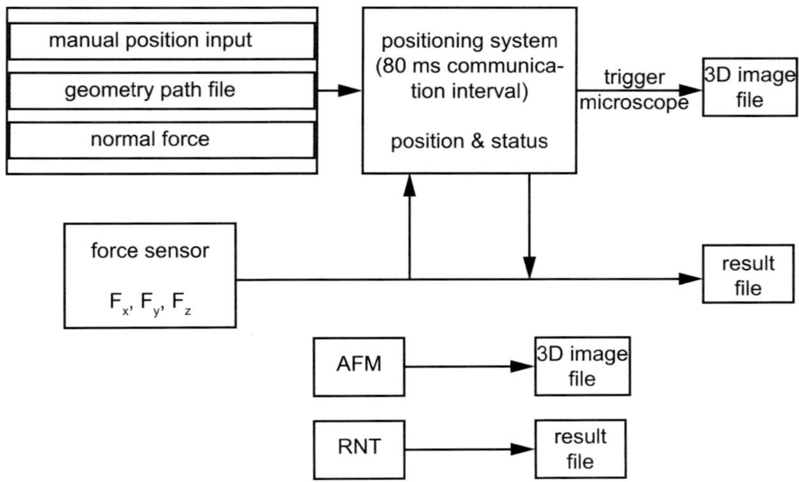

Bild 2.13.: Software and control diagram.

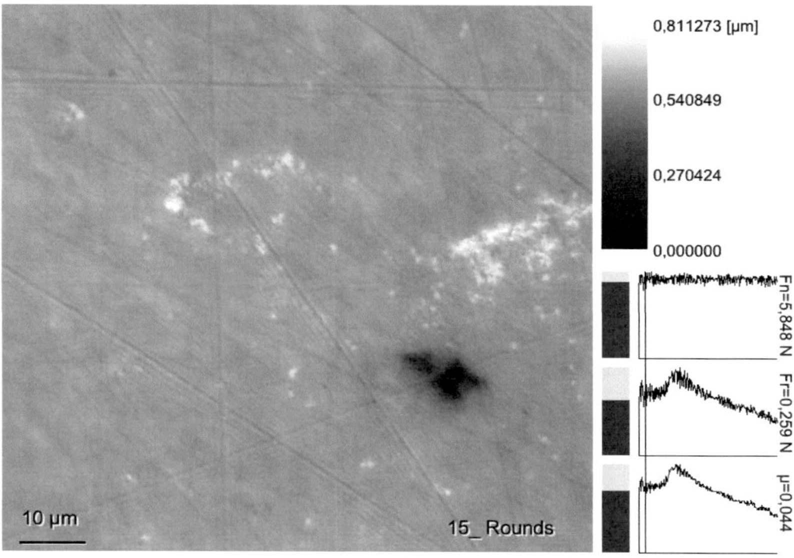

Bild 2.14.: Representation of 3D data from a Cu sample surface (left) with a grey scale for the z-axis (top right) as well as measured forces and friction coefficient for the current cycle (lower right).

struments Corporation) program. The connection to the positioning stage is established via TCP. A NI PXI-6120 S Series card (National Instruments Corporation) records the three analog signals (x, y and z) of the force sensor. Separate computers control the optical and holographic microscope as well as the RNT. A TCP connection establishes the communication to the computer which is directly attached to the holographic microscope. This enables the remote control of the microscope and consequently the automation of the acquisition.

The geometry of the sliding path of the planar drive is programmed in a G-code similar manner. The current position is read every 80 ms. The z-position is adjusted according to the current value in order to maintain the normal force constant. Once the objective lens of the holographic microscope has reached a predefined position, the stage pauses and the acquisition of a 3D image is triggered.

3D images gathered during the experiment are organized in frames, which provide an animated view of the surface and graphical representation of forces and friction coefficients. The values shown at these graphs are calculated via interpolation for the position examined with the microscope. The two closest measured samples to this position are used for the interpolation. An example is shown in Figure 2.14.

Software structure

The software comprises two main loops. The first one is assigned to gather the analog output of the FOS. The output of the FOS module is a voltage value which is not linearly proportional to the distance between the sensor and the mirror. Therefore, the raw signal is translated to distance according to voltage/distance curves provided for each FOS unit by the manufacturer. The sensors used in this instrument have two measurement regions, near and far, depending on the distance of the FOS from the mirror. Both can be determined by the calibration curve. Although the near region is more precise, drifting and external factors render this region too complicated to use in this setup. A local file contains the calibration data for each force sensor (k_x and k_z) and measurement region, which are used to convert the distance to forces, provided that the sensor's deformation is linear and elastic. Friction coefficient values can be calculated from the measured forces ($\mu = F_F / F_N$).

A library of commands was built for the positioning system according to the API (application programming interface) provided by the manufacturer. These commands consist of basic functions such as defining a new position for manual movement or reading the current position and instrument status. Part of this library involved converting all variables from little to big endian in order to establish communication between LabView and the positioning system.

All commands were organized in subroutines. The main program can call them to perform various functions, such as loading curves, starting a curve, deactivating air-pressure, etc. A function worth mentioning is the normal force regulation, which uses the currently measured normal force and retracts or approaches the pin with variable speed according to its deviation from the preset value. The software reads the current status of the positioning system at the end of every cycle of a main loop. While transmitting the input commands given by the user or an automated operation, all other modules, including status reports, are paused. The same holds when sending or receiving command related data.

An additional library was built to include all xml commands used for the remote control of the holographic microscope. The most important functions used for this project are acquiring a double wavelength holographic image and saving the tif image as detected by the CCD sensor. Additional commands open a tif image and export the height information into an ASCII file containing the height calculated for every pixel.

The user-interface consists of a main panel with all indicators of the current position (x, y and height of force sensor) and the forces currently measured. A 2D plot shows the current position of the pin, as well as the path that has been programmed and what part of it has already been followed. A second tab displays the currently exported image from the microscope. In the same panel, there are buttons for main functions, such as starting or aborting an experiment, pausing, displaying the position and regulating the normal force.

The second half of the user interface comprises tabs with gathered controls:

- Startup: contains a set of buttons that initiate the hardware and start the microscope remote control module.

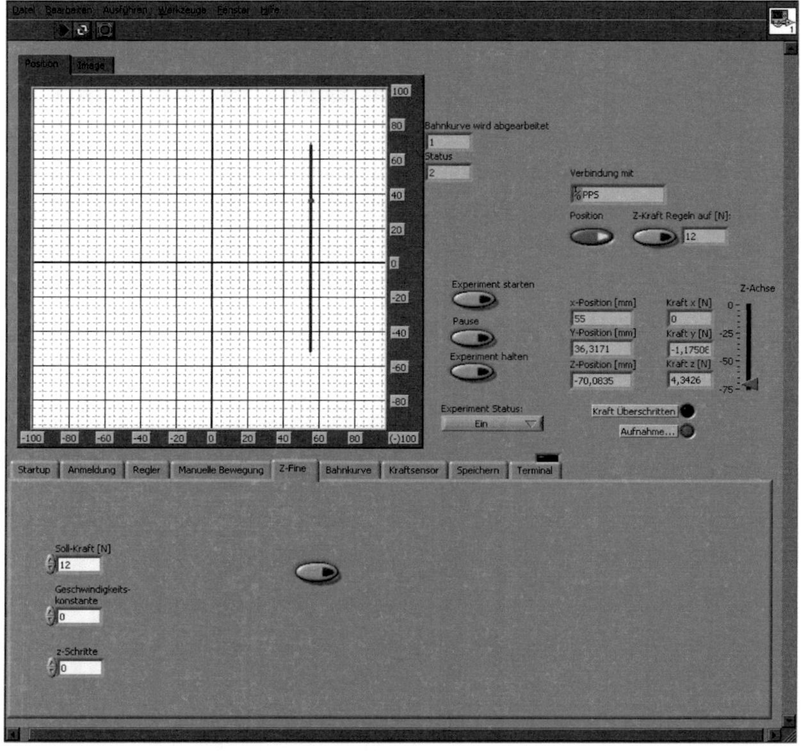

Bild 2.15.: Snapshot of the user-interface while a linear reciprocating motion experiment is in progress.

- Login: activates the internal software of the positioning system.

- Regulator: activates air pressure and begins the position calibration.

- Manual position: includes arrow buttons, speed settings and manual input of target position.

- Z fine regulation: contains the input values for the force regulation.

- Curve setup: reads the path of the file containing the curve coordinates and speeds, as well as a set of commands to start/stop a curve.

- Force sensor: contains buttons to tare the forces and a display with the current distance values between FOS sensors and mirrors.

- Save: has options such as the path and file names for the exported data (forces and images) are defined.

- Terminal: displays all messages received from the positioning system (including error messages).

A snapshot of the user interface is shown in Figure 2.15

2.3. Calibration and Testing

2.3.1. AFM

As the positioning system readjusts its position dynamically during operation, fluctuations of a constant position may occur. These may affect the stability of the AFM measurements. To ensure the stable function of the atomic force microscope, several tests were performed with dry and water immersed samples. An example of a Cu surface is shown in Figure 2.16. The surface was examined with the positioning system turned off, as well as with the positioning system turned on and the AFM working in both dry and immersion mode. Given that no additional shielding or damping was applied to improve the scanning process, the resulting images have acceptable levels of noise and the surface structure is well recognizable.

2.3.2. Position calibration

In order to examine a single position on a flat sample with all three sensors (force sensor, holographic microscope and AFM), their relevant position has to be determined. This task was performed in two steps. First, a sphere was attached to the force sensor, in the same position where the tablet shaped sample is normaly located. With this setup, a cross was engraved on the plate sample at a given position of the stage (x_0,y_0). The

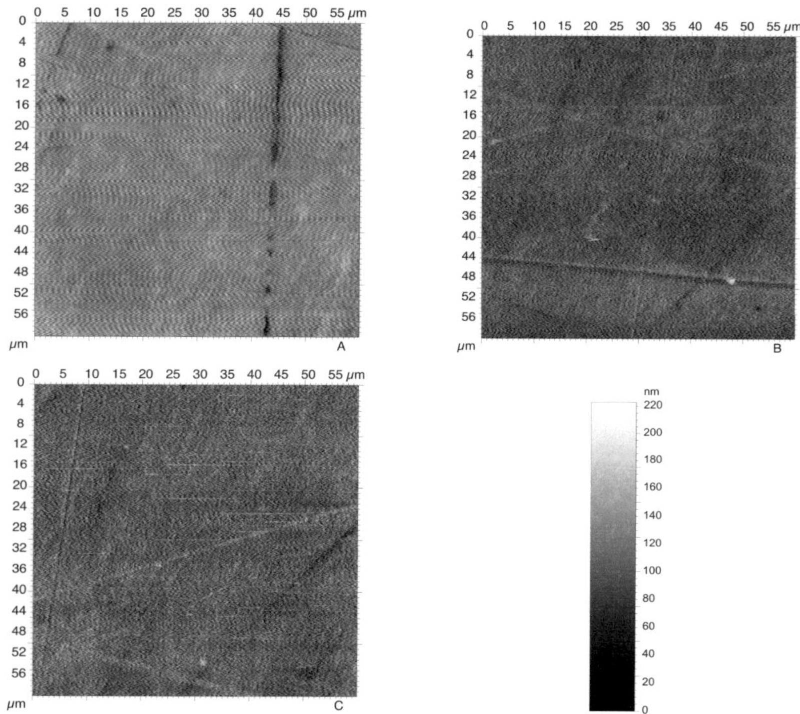

Bild 2.16.: AFM measurements of a polished copper surface in dry mode with the positioning system off (A), in dry mode with the positioning system on (B), and in water immersion with the positioning system on (C).

center of this cross was then located with the holographic microscope by moving the stage to a new position (x_1,y_1). A patterned silicon sample with numbered blocks was then placed on the same position on the plate sample and an image was acquired with the holographic microscope. The stage was moved again until the same position of the patterned sample was relocated with the AFM (x_2,y_2). Images of the patterned sample are shown in Figure 2.17, which demonstrates the precision with which the instruments can be synchronized.

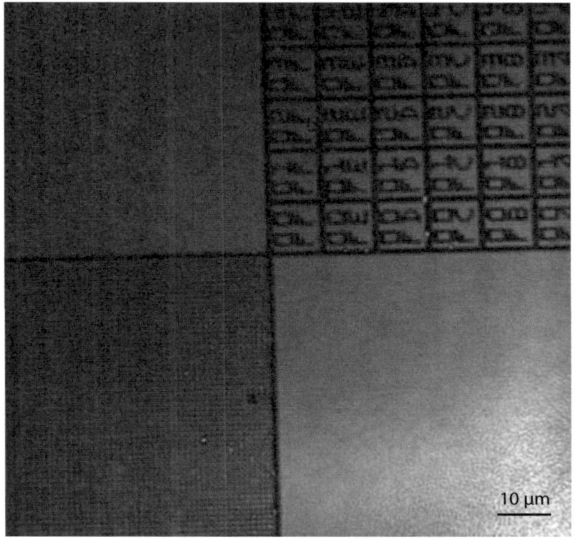

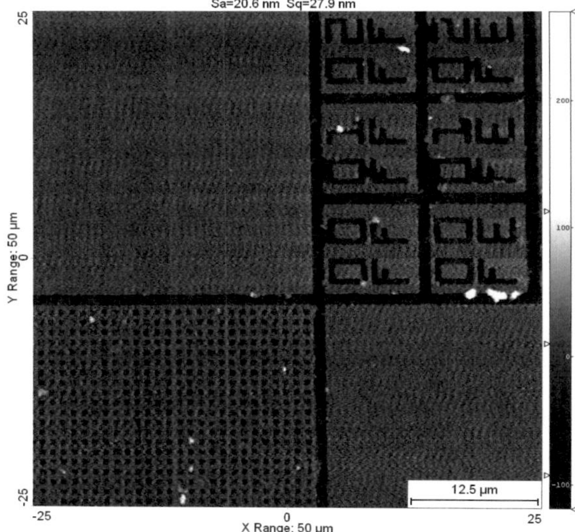

Bild 2.17.: Reference patterned sample with numbered blocks located with the holographic microscope (top) and the same sample relocated with the AFM (bottom) by moving the stage.

When designing a path for a tribological experiment it is necessary to program a path that is accessible to the force sensor and the holographic microscope. The simplest case is a line that connects the relative positions of these elements. The slope and offset of this line is calculated as shown below,

$$A = \frac{y_1 - y_0}{x_1 - x_0},\tag{2.2}$$

$$B = y_0 - Ax_0.\tag{2.3}$$

In such an experiment, if the pin is in contact with the tablet at position (x_a, y_a), which is to be examined, then the acquisition with the holographic microscope will be made at (x_b, y_b):

$$y_b = y_a + y_1 - y_0,\tag{2.4}$$

$$x_b = x_a + x_1 - x_0\tag{2.5}$$

and for any other position (x, y) a random y position is coupled to

$$x = \frac{y - (y_a - Ax_a)}{A}.\tag{2.6}$$

The path is further adjusted to include the position of the stage for an AFM measurement.

2.3.3. Example of an experiment

An extended test was performed to try the instrument in experimental conditions. For this purpose, two samples of pure iron (>99.9 % purchased by Goodfellow GmbH, Germany) were prepared: a tablet and a plate shaped sample. The edges of the tablet were tapered to avoid indentation of the sample's side in the counter-face and the contact surface was polished. The contact area was round in shape with a diameter of approximately 1.9 mm.

The sample was labeled using neutrons in a nuclear reactor. The specific activity of the sample was 17.21 kBq (Fe-59).

A linear path connecting the force sensor and the holographic microscope was programmed. The stage moved in a reciprocating motion at 15 mm/s. The normal force was set to 10 N (nominal pressure of 2.7 MPa) and the length of the line was 120.03 mm. A total of 5450 cycles were recorded, which corresponds to a total sliding length of 654.17 m. The experiment was performed using poly alpha olefin as lubricant.

During the first 100 cycles, the precision in the positioning and force regulation were monitored. Since the programmed path was a straight line, the deviation of the stage position from the theoretical line was calculated by their perpendicular distance. The results are shown in Figure 2.18. The average deviation calculated was 18 nm. This result proves that the path is followed precisely enough to ensure that the pin remains on the same track throughout the test. This deviation does not affect the acquisition of the 3D images, as it is lower than the resolution that the holographic microscope provides. On the other hand, the dynamic stabilization of the normal force showed greater deviations. As the plate sample was not polished, the roughness of the surface had an impact on the normal force regulation, which gave rise to a standard deviation of 31 % of the set force.

Figure 2.19 gives an overview of the entire experiment. In the beginning, both the friction coefficient and the wear rate increase. In this phase of the experiment, no significant change is observed in the images acquired by the holographic microscope. After 600 cycles, the left part of the image starts changing. This happens gradually, and expands to further asperities coming into contact in the rest of the window observed with the microscope. In this phase, the friction coefficient and the wear rate start to stabilize. Deeper scratches in directions other than the sliding direction slowly fade away. After 1200 min these scratches are no longer recognizable and the wear scar has expanded on the entire surface. A total value of 786.5 nm wear was calculated as total wear at the end of the experiment.

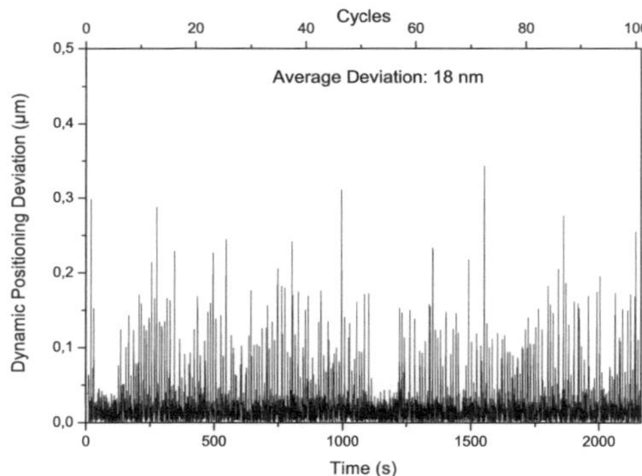

Bild 2.18.: Dynamic perpendicular positioning deviation of the stage during the first 100 cycles of a reciprocal motion with a linear path of 120 mm. The experiment had a set normal force of 10 N with a sliding speed of 15 mm/s of an iron pin against an iron plate with PAO lubrication.

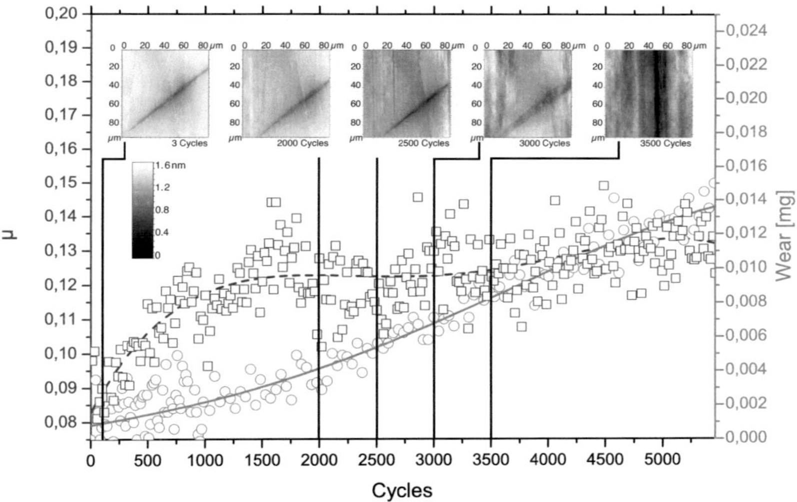

Bild 2.19.: Evolution of the topography, friction coefficient (open squares) and measured wear with RNT (open circles) during frictional load (normal force) of 10 N and sliding speed of 15 mm/s of an iron pin against an iron plate with PAO lubrication.

It is interesting to note, that the wear rate and topography changes seem to be correlated, while the friction force evolves independently.

2.4. Summary

The presented tribometer combines modern techniques that enable the monitoring of the surface of one of the two tribological counter-faces, while measuring normal, frictional forces and wear. Its major breakthrough is the fact that its function is not limited to dry or transparent systems. On-line topography in lubricated tribosystems can help understand the phenomena that occur in several applications in greater depth.

Thorough tests have been performed to determine the precision and stability of function. These show, that one single position of the wear scar can be relocated and examined after each cycle of an experiment. This allows for dynamic observations of the changes that occur. The resolution offered by the holographic microscope is sufficient to provide information on the plastic deformation that occurs at specific asperity junctions. Further detail can be obtained by scanning the same position with the AFM. At the same time, the behavior of the friction coefficient and wear rate can be displayed throughout the experiment. As these usually exhibit a non-linear behavior against time, interesting changes can be coupled with observations on the surface of the sample.

In this chapter, the results of only one simplified experiment with a linear path and iron samples were presented. Nevertheless, this tribometer can be utilized for experiments that demand greater geometric complexity. In addition, as there are few requirements for the materials of the selected samples, several material combinations can be tested. One important limitation is connected to the on-line wear measurement: the samples must contain isotopes that can be activated.

3. *In-situ* observation of wear particle formation on lubricated sliding surfaces

3.1. Introduction

Since the first studies of wear phenomena, a wide variety of terms have been used to describe and categorize wear processes. These often depend on the background of the research and the particular effects being studied. Even when delving deeper into subcategories of wear, there has been controversy as to what words better describe the experimental observations. A chapter in the theory of wear that received a lot of attention in the 1970s was that of delamination in sliding, fretting, rolling, and grinding. There has been considerable debate over the validity of the delamination concept. The focus of the present work is on the dynamics of wear phenomena on lubricated Cu sliding surfaces, especially those that are often interpreted as delamination without sufficient experimental proof. In order to better describe wear mechanisms, single events were tracked by *in-situ* optical topography measurements.

The delamination theory of wear was first introduced by N.P. Suh in the early seventies [34, 35]. His approach aimed at suggesting answers to questions that arose from reports of flake-like sheets as well as lamellar structures and sub-surface cracks. The theory is based on the piling-up of dislocations under the surface, leading to voids that coalesce and form cracks parallel to the surface. The size of these cracks depends on the sub-surface structure of the material. Once they reach weak positions, they shear to the surface causing long and thin wear sheets to "delaminate". An example of such weak positions is a pre-existing crack, very often found under the oxide layers of Cu.

A systematic research and reviewing of relevant work was conducted by D.A. Rigney et al. [36, 37, 38, 39]. In their work, several wear mechanisms are compared and the parameters that affect sliding wear are analyzed. However, it should be emphasized, that many of these mechanisms, including delamination, are mostly studied in dry systems.

One of the strongest tools to identify the wear mechanisms at play have been cross-sections at the worn surface. Unfortunately this method cannot clarify the question of surface vs. sub-surface origins for debris-producing fracture. As argued in a relevant review by Rigney [36], a cross-section is a destructive method, that provides no dynamic measurements, and can only extract information from two-dimensional slices. Most importantly, other processes occurring on the surface, such as transfer, mixing, folding, and environmental effects could be responsible for features which mimic sub-surface cracks. The author stresses the importance of a non-destructive technique with adequate resolution and response time in order to solve the aforementioned problems.

Although simulations can be very helpful in suggesting answers to questions regarding wear under cyclic loading, they need to be confirmed by experimental results. Recently, important breakthroughs have paved the way to revolutionary experimental investigations of sliding surfaces.

A notable number of *in-situ* tribometers have been reported as suitable means to monitor the contact area [31], buried interface, transfer films [40] and the evolution of wear tracks. These exhibit various advantages and disadvantages. There is no universal solution that can simultaneously analyze all aspects of sliding surfaces. Certain limitations need to be taken into consideration, such as the use of a transparent counter-face, testing in air or vacuum [41] etc. The instrumentation used in this work allows for simultaneous measurement of surface topography of a lubricated sliding surface in immersion and is described in detail in chapter 2.

3.2. Experimental Procedure

All friction/wear experiments were conducted in laboratory conditions: temperature, $T=25\,°C$, humidity, approx. 45 %. *In-situ* topography images were acquired after every loading cycle with a holographic microscope (DHM® R1000 series, Lyncée Tec SA). As the objective lens remained in immersion during acquisition, a transparent PAO-8 oil with a viscosity of $45.8\,mm^2\,s^{-1}$ at $40\,°C$, provided by Fuchs Petrolub AG, was selected as lubricant. The oil circulation was performed by a pump integrated in the radio nuclide technique apparatus (Zyklotron RTM 2000 of Zyklotron AG, Germany). During the experiment, the lubricant's temperature was stabilized at approx. $33\,°C$. The plate samples were made of Cu (pure 99.98 wt%) and polished with a $1\,\mu m$ diamond particle suspension without further treatment. The average grain size measured on the surface was $1452\,\mu m^2 \pm 71\,\mu m^2$. The pins were made of Fe (pure 99.98 wt%) and bearing steel (1.3505 / 0.9-1.05 % C and 1.35-1.65 % Cr). The contact area was circular in shape with a diameter ranging from $1.6\,mm$ to $2\,mm$ depending on the edge taper. All samples were polished before testing in the same way as the Cu plates. The pin samples had a slight curvature due to the polishing procedure. The measured root mean square roughness (rms) was measured for both sample types. For the Fe samples the rms roughness was approx. $50\,nm$ and for the steel samples approx. $200\,nm$. The roughness was recalculated after subtracting a spherical term in order to eliminate the curvature factor, leading to values of approx. $10\,nm$ for Fe and approx. $26\,nm$ for the steel samples.

Experiments of two different types of sliding paths were conducted: in a linear reciprocating motion and along a parallelogram with rounded edges, which resulted in a uni-directional load. Both are illustrated in Figure. 3.1. The nominal pressure for the experiments conducted for the present work ranged from 1.5 to 2.5 MPa and the sliding speed from 10-20 mm/s, de-

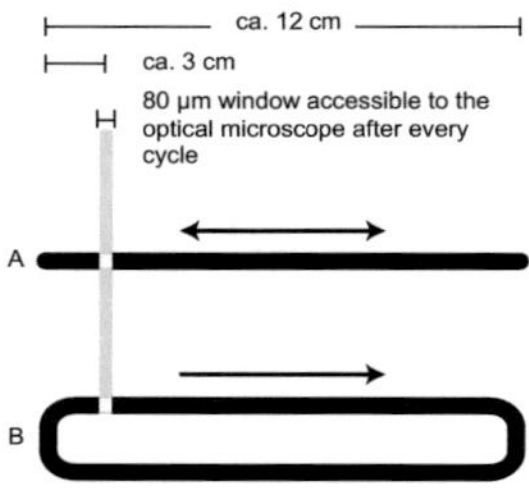

Bild 3.1.: Wear track geometry for linear reciprocal (A) and curved parallelogram/uni-directional (B). The white squares in the gray area indicate the position of the $80\times80\,\mu m^2$ window viewed by the optical microscope.

pending on the experiment. The exact conditions are presented in the results section.

All holographic 3D images were recorded at the same spot of each wear track, allowing for dynamical observations of the topography evolution of the wear track. Normal (F_N) and frictional (F_F) forces were calculated via interpolation for the exact position that was examined with the microscope. In the case of a reciprocating motion, the pin comes into contact with this position twice during every cycle, once in each sliding direction. For the calculation of the friction coefficient (μ), F_F was averaged from the values of each direction.

Selected samples were examined after the tribological experiment with additional methods. Chemical analysis of the samples was performed with an XPS system (PHI 5000 VersaProbe[TM], Physical Electronics Inc.). After obtaining spectra from the surface of the samples, an ion gun was employed for depth profiles. A Zeiss XB 1540 FIB was used to investigate the structure of the samples underneath the worn areas. In order to observe the depth profile, cross-sections along and perpendicular to the wear tracks were milled with a Ga ion gun. No mask was required for the purposes of this analysis, as no thin slices for transmission electron microscopy (TEM) investigations were produced. The images were acquired with the secondary electron detector.

3.3. Results

3.3.1. Evolution of the friction coefficient

Generation of flake-like particles was observed in most, but not all experiments. Even without altering the experimental conditions between tests, μ often remained at values below 0.04 and no significant topography changes occurred. In the experiments that were further studied in this work, the initial μ was low until an abrupt increase, which triggered profound changes in the topography and an increase in surface roughness accompanied by wear. After several cycles, μ decreased again and stabilized at a slightly higher value than the one recorded during the first cycles. The aforementioned abrupt increase of μ was necessary to study the processes that lead to the generation of wear particles, so F_N was initially increased, regulated stepwise and was then reduced to the preset value for the experiment. A representative evolution of μ as cycles progress is shown in Figure 3.2. At 4192 cycles the experiment was paused before readjusting the microscope settings and dropping the normal force from 6 to 4 N. The slider remained in contact for about 4 min. Within 50 cycles after resuming the experiment, the friction force became unstable and increased at about 4220 cycles. The data gathered around the values of 4192 and 4220 cycles are insufficient to locate the exact moment that the friction values showed spikes, indicating the beginning of drastic changes on the surface. It is noteworthy, that for this experiment, the acquisition of force values was performed every 300 ms and the field of view was limited to about $80 \times 80\,\mu m^2$. The surface of the sample remained rough after running in. The average roughness (Ra) of the polished Cu surface was 183 nm, after 5000 cycles 614 nm and after 8000 cycles 520 nm.

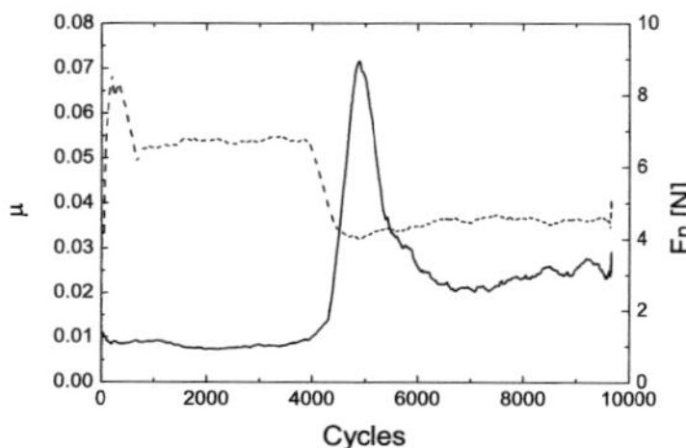

Bild 3.2.: Control of normal force (dashed line) and evolution of friction coefficient (continuous line) of a running in experiment with a preload phase. An average smoothing filter was applied to both lines, in order to remove the high frequency part of the forces. Steel pin sliding against Cu with a speed of 20 mm/s.

Particle generation in the early wear stages, shortly after the increase of F_F, are the simplest to monitor, because the surfaces are still relatively flat and worn areas are easier to distinguish from intact areas within the wear track. This feature is related to the method of holographic microscopy. A high concentration of steep slopes increases the possibility of 3D reconstruction errors. Therefore, the results that follow are mostly gathered in these stages.

3.3.2. Occurrence of local wear

Local wear was investigated with topography images acquired with a holographic microscope *in-situ* and SEM images after the experiment. To ensure that the same kind of structures were studied with both instruments, images of wear-induced pit sites were obtained at comparable magnifications with both methods. It can be seen in Figure 3.3, that the size and shape of the pits in either case are very similar. In both images, the length of the detachment sites was of the order of 10-20 µm and the width was about half the length. Grooves ran along the sliding direction and wavy patterns were formed, again in the sliding direction.

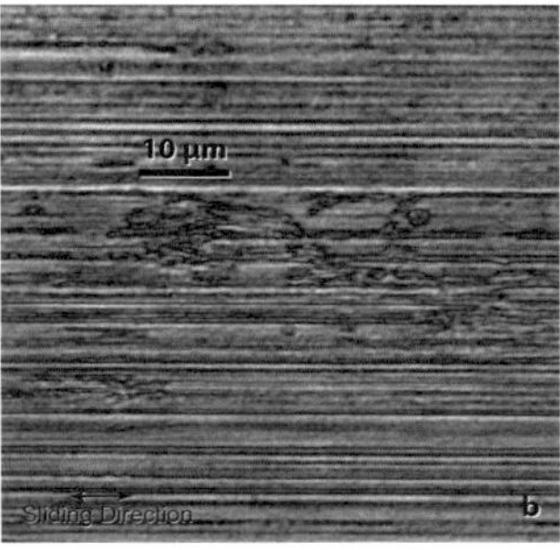

Bild 3.3.: Delamination on the Cu surface recorded *ex-situ* with a SEM (a) and *in-situ* with the holographic microscope (b). In the case of the *in-situ* image, the objective lense was immersed in PAO-8 during acquisition. The picures show the electron and light intensity images respectively. Steel pins slid in a linear reciprocal motion against both samples with a nominal pressure of 1.5 MPa and a speed of 20 mm/s.

Each method was further used to delve into different aspects of wear generation. SEM images of FIB cross-sections shed light on the processes that occurred during sliding underneath the surface, while the holographic microscope recorded successive topography images of the surface, allowing an observation of how wear evolves.

3.3.3. SEM images and cross-sections

A polished surface was first examined, in order to obtain a reference. The result is shown in Figure 3.4a. The ion milling reached a 7 µm depth. The grains were too large to fit fully into the field-of-view, with the exception of a smaller grain structure near the surface, which did not exceed 1 µm in depth. No nano-crystalline (nc) or ultra-fine-crystalline (ufc) structures were observed directly underneath the surface. Entirely different results can be seen in Figure 3.4b. This is the result of 9743 cycles of sliding. The grains are significantly smaller and appear to be elongated along the sliding direction. Their structure becomes finer and finer when approaching the

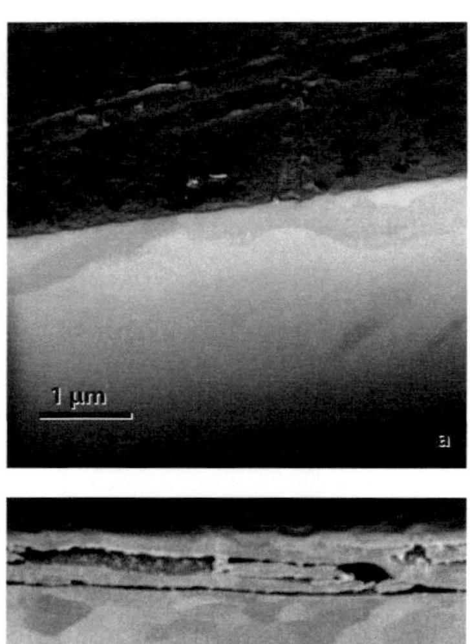
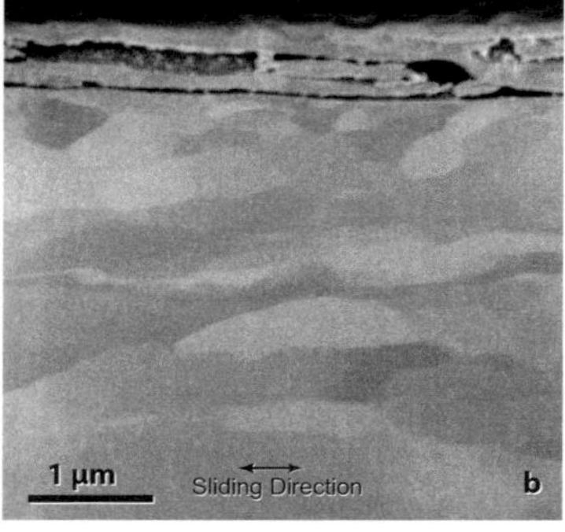

Bild 3.4.: Comparison of the depth profile of a polished Cu surface (a) to that of a wear track after a linear reciprocal tribological load for 9743 cycles with a nominal pressure of 1.5 MPa and a speed of 20 mm/s (b). The cross-section was milled along the sliding direction. The sample stage was tilted by 54° during acquisition.

surface, reaching diameters below 100 nm. A lamellar thin layer of about 300 nm forms the surface of the sample. Here, only nc and ufc structures can be found. In the example of Figure 3.4b the lamellae appear to be practically separated from the underlying grains. The dark areas within the fine structure are voids.

The above result corresponds to an experiment in which the motion of the slider was reciprocating. A slight difference was noticed when the surface

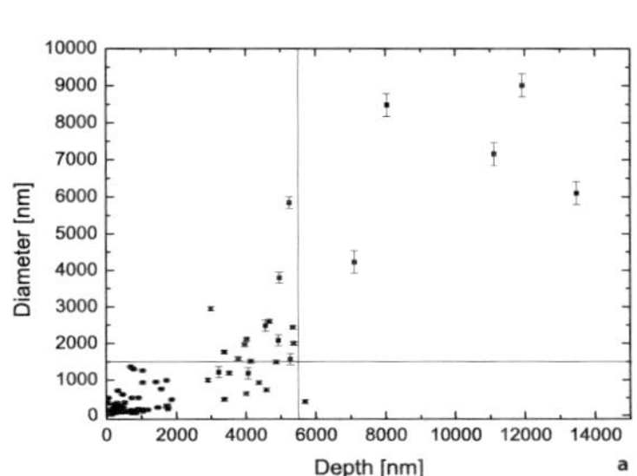
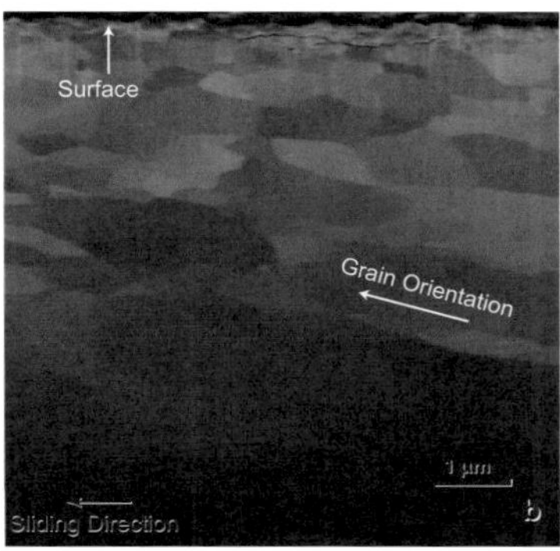

Bild 3.5.: Sub-surface grain structure. The graph on the left (a) shows the size distribution in depth and the image on the right (b) is a FIB cross-section of a Cu sample that was only loaded in one sliding direction for 5500 cycles with a nominal pressure of 2 MPa and a speed of 20 mm/s. The cross-section was milled along the sliding direction.The sample stage was tilted by 54° during acquisition; therefore the depth values were corrected.

was loaded in one direction only: the grains have a morphological orientation toward the sliding direction. This is demonstrated in Figure 3.5b. In Figure 3.5a, the size of the grains are plotted against depth. A large number of nc and ufc grains are located near the surface. The length of the grains increases gradually from sub-micron to about 9 µm at a depth of 14 µm.

Another feature observed in FIB cross-sections were vortex-like structures. The clearest example from these measurements is shown in Figure 3.6. Such formations were not found extensively in the sample; however, it is worth mentioning that the subsurface structure that was formed by vortices resembles the lamellar structures studied in this work.

As stated above, in many experiments, the anticipated abrupt increase of μ did not occur. One of the samples involved in these experiments was examined as well. The result is shown in Figure 3.7. The grain structure resembles the sub-surface of the pristine, polished Cu sample.

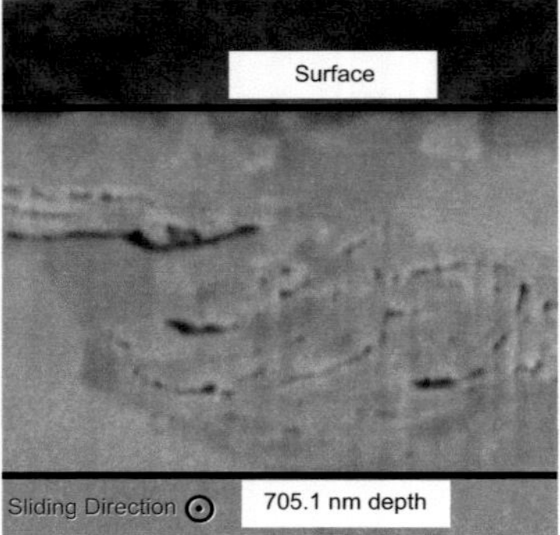

Bild 3.6.: FIB cross-section of a vortex-like structure in a Cu sample that was only loaded in one sliding direction for 5500 cycles with a nominal pressure of 2 MPa and a speed of 20 mm/s. The cross-section was milled perpendicular to the sliding direction.The sample stage was tilted by 54 ° during acquisition, therefore the depth values were corrected.

Bild 3.7.: Depth profile of a Cu surface after a linear reciprocal sliding for 3775 cycles with a nominal pressure of 2.5 MPa and a speed of 20 mm/s. The experiment was stopped at an early stage, before noticing an increase in friction. The cross-section was milled along the sliding direction. The sample stage was tilted by 54 ° during acquisition.

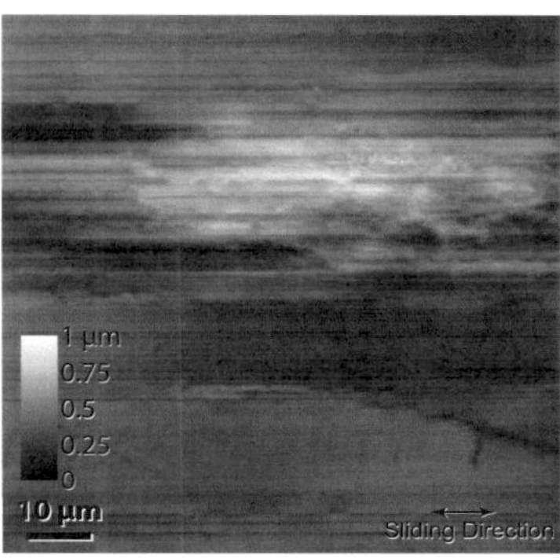

Bild 3.8.: Topography image of the Cu surface during the experiment. A greyscale bar on the lower left indicates the height. A steel pin slid in a linear reciprocal motion against the Cu sample for 5072 cycles with a nominal pressure of 1.5 MPa and a speed of 20 mm/s.

3.3.4. Propagation of damage zone on the surface

While the diameter of the pins was in the mm range, the area that was observed with the mircoscope was $90\,\mu\text{m} \times 90\,\mu\text{m}$, in order to monitor the microstructural changes on the surface. Although both samples were polished, the spreading of worn zones within the apparent contact area occurred gradually. In the beginning only small strips appeared along the sliding direction. As the experiment continued some of these strips grew wider and deeper. Even after several cycles, some areas remained intact. The positions where particles were detached rose higher shortly before wear took place. This is shown in Figure 3.8. The brighter area is an elevated structure where wear particles detached within 100 cycles after acquiring this image. The horizontal strips seen in a large part of the picture are parallel to the sliding direction. There is a small area that runs through the image in the lower part, where minor scratches appear in directions other than that of sliding. These scratches originate from the polishing procedure, so this area has not undergone any plastic deformation yet.

The evolution of the elevated area was monitored as it spread along the sliding direction. Figure 3.9 shows the progress of a zone where particle detachment occurs. In the beginning, this formation is limited to the right side of the image. Brighter objects indicate particles on the surface. The elevated structure appears to spread to the left in an oval shape. It keeps elongating until the end is outside the field-of-view. Dark stains within this area correspond to pits. It appears that flakes can still remain on the surface covering the greatest part of the pits. This is perhaps easier to recognize in the intensity images of the holographic microscope, as demonstrated in Figure 3.10a. Figure 3.10b shows the height information derived from a profile along the sliding direction at the location of the particle, whereas Figure 3.10c is a FIB cross section demonstrating a similar flake-like particle forming within the lamellar structure. Although Figure 3.10c does not correspond to the same flake, it shows a similar structure.

To get a better picture of this progress, the spreading was measured from the moment the elevation appeared on the right of the image. This is displayed in Figure 3.11. The elevated area does not appear to propagate linearly, although the rate is almost constant from 5060 to 5078 and from 5078 to 5092 cycles. In this experiment, no particular topography change indicates a transition at 5078 that could affect the spreading rate.

Similar results were recorded for experiments of pure Fe sliding against Cu. In one of these tests, the displacement and spreading of a single asperity leading to the formation of a pit was observed. Profiles of the surface were drawn from the topography images along the sliding direction. Figure 3.12 shows the profile at the same position after 606, 615 and 626 cycles. At 606 cycles, a large asperity that stands out from the rest of the surface can be seen in the profile. In cycles 615 and 626, the volume of this asperity appears to have moved to either side of the initial position, but mostly to the right. A pit has been formed in the initial position. As the surrounding area constantly changes, it is impossible to measure the exact volume smeared

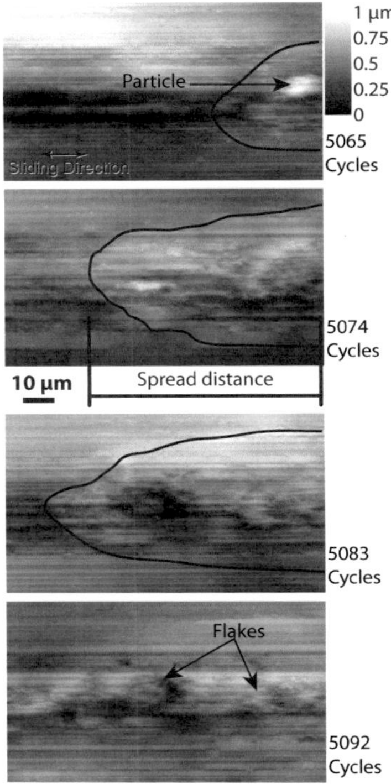

Bild 3.9.: Progress of an elevated area exhibiting particle generation events recorded with the holographic microscope during the sliding experiment. The greyscale bar on the top right of the image represents the height in the 3D images. The black curved lines indicate this area as it spreads along the sliding direction. Particles and flakes appear on the surface throughout this process. A steel pin slid in a linear reciprocal motion against the Cu sample with a nominal pressure of 1.5 MPa and a speed of 20 mm/s.

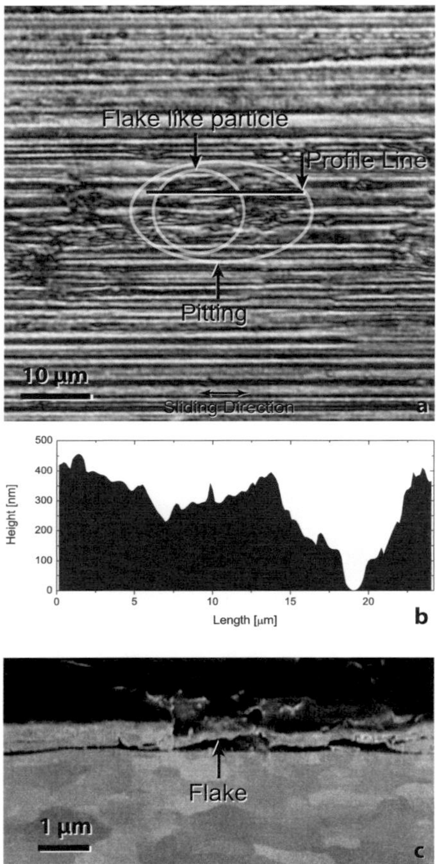

Bild 3.10.: Intensity image of the holographic microscope (a) and derived height profile (b) showing a flake-like particle hanging over a pit. A steel pin slid in a linear reciprocal motion against the Cu sample with a nominal pressure of 1.5 MPa and a speed of 20 mm/s. FIB cross section of a flake-like particle (c). For comparison purposes, the image of Figure c was mirrored.

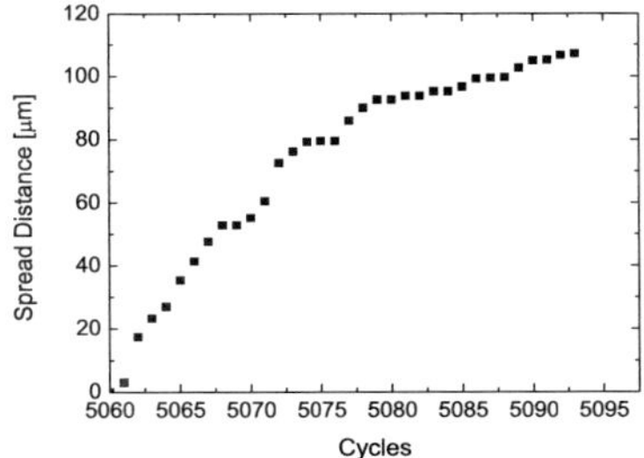

Bild 3.11.: Spread distance and rate of the elevated area exhibiting multiple delamination events. A steel pin slid in a linear reciprocal motion against the Cu sample with a nominal pressure of 1.5 MPa and a speed of 20 mm/s.

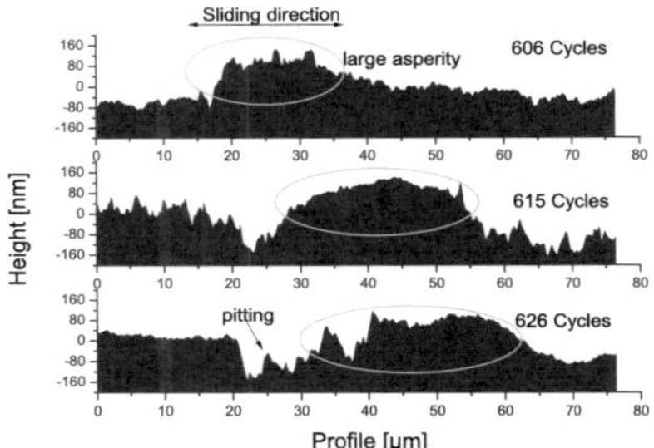

Bild 3.12.: Profiles along the sliding direction showing the progress of a single asperity moving along the sliding direction, leaving a pit behind. A Fe pin slid in a linear reciprocal motion against the Cu sample with a nominal pressure of 2 MPa and a speed of 10 mm/s.

around the initial position of the asperity and the exact volume lost in the pit.

3.3.5. XPS investigation

To investigate possible transfer to the counterface, steel pins and Cu samples that had slid against each other were examined by XPS. After testing several pins with the tribometer, stains on the pins' surfaces were visible to the naked eye. A steel sample with larger stained areas was selected to determine the chemical composition of the surface. Peaks of C, O, Fe were dominant in the spectra acquired on the surface. The surface appeared to be covered by iron oxide and carbon or organic material. Apart from these peaks, Cu also appeared to be present, so the sample was sputtered in order to obtain a depth profile. The sputter rate was calibrated for this instrument using SiO_2, yielding 2nm/min of SiO_2. The peaks monitored are shown in the legend of Figure 3.13, which shows the depth profile. Oxygen and organic material dominated the surface. Underneath, Cu appears to rise in concentration particularly fast. The highest concentration is detected after 30 min of sputtering. This corresponds to more than 60 nm of depth. As discussed in the following section, this observation is related to material transfer. At the same depth, the concentration of organic material has decreased dramatically. Oxygen has also dropped, but remains at high levels compared to its surface concentration. The Cr peak was hardly visible throughout the test. Cu remains at a concentration over 10 at.% even at depths greater than 300 nm.

Considering the interesting lamellar structure revealed by the FIB cross sections, depth profiles were also performed on the Cu surface. Figure 3.14 presents a depth profile acquired at an intact area of the sample and another one acquired from the worn surface. It is noteworthy, that the sputter rate here was set to 6 nm/min. O and C were detected both in areas of the surface that did not come into contact with the slider, as well as in the wear track. A significantly higher concentration of these elements was measured on the worn surface. On the pristine surface the atomic concentrations of O and C were 19 and 27 at.% and on the worn surface 34 and 61 at.% respectively. In the depth profile of the intact surface, after a couple of minutes of sputtering, the Cu peak dominated the spectrum. This means, that below a depth of 20-30 nm the sample consists of pure Cu. This is not the case under the wear scar, where the peaks of O were still distinguishable even after 40 min of sputtering.

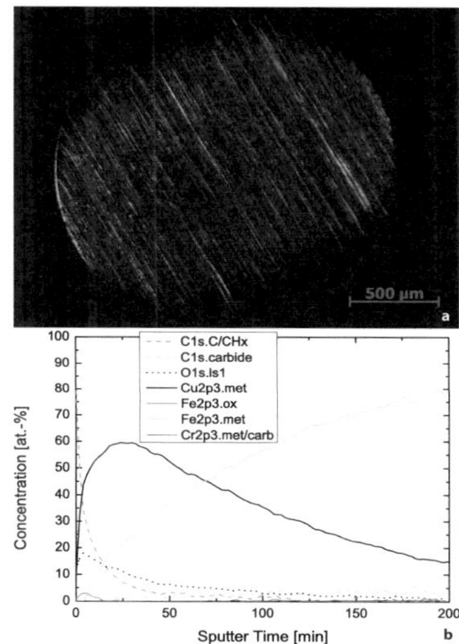

Bild 3.13.: Surface (a) and XPS depth profile of a steel sample (b) that slid against a Cu sample in a reciprocal motion for 3775 cycles with a nominal pressure of 2.1 MPa and a speed of 20 mm/s.

3.4. Discussion

Polishing the Cu sample only affected a thin layer near the surface. Grains deeper, toward the bulk of the material, were significantly larger. The diamond particles used in the polishing suspension are not mixed in the material. This is evident in the XPS depth profiles, as 1 µm particles would have given C peaks at depths far below the thin 20-30 nm layer detected.

During the first cycles of all experiments, low μ values were recorded. The low initial roughness of the Cu surface cannot account for this observation, as the roughness of the surface after running-in did not drop to the levels of the polished surface. The oxide layer on the surface of the polished sample appears to provide wear resistance in the beginning of the experiment.

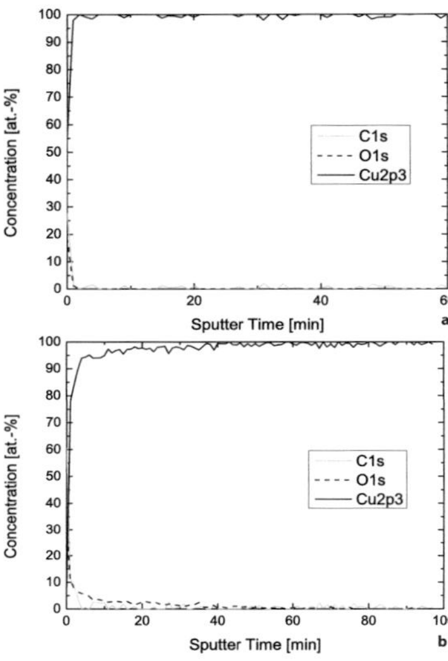

Bild 3.14.: XPS depth profile of a polished Cu surface (a) and a Cu sample that was loaded in one sliding direction for 5500 cycles with a nominal pressure of 2 MPa and a speed of 20 mm/s (b).

With the data currently available, the abrupt increase of μ can only be attributed to random asperity contacts leading to the breach of the thin oxide layer and local wear. Further experiments with increased time resolution and a wider field of view are necessary to shed further light on the initiation of severe wear and increase in F_F. Due to storage limitations of the current experimental setup, the force values cannot be recorded at such high rates when performing an experiment with thousands of cycles. Fragmentation of the result data, or setting criteria for the adjustment of the acquisition rate according to the instability of F_F can be applied in future work.

A running-in phase lasts about 1000 cycles, until the friction coefficient stabilizes below 0.03. Besides the consideration of oxide layers, a transition of wear regime from local damage to delamination of a mechanically mixed layer has been reported for increased normal forces on nanocrystalline surface layers of pure Cu [42].

Dents, such as those presented in Figure 3.3, are often reported as delamination wear [43, 44]. Rigney et al. have differentiated flake-like wear debris produced by delamination from non-flake debris by lamellar extrusions [37].

The grain structure underneath the wear track changed profoundly during the cyclic load. Apart from the finer grain structure, there was a 300 nm lamellar structure directly underneath the surface. The hardness and ductility of this zone is expected to differ from the polished surface [45, 46, 47]. It appears that the formation of flakes and the wear particle generation occurs in this layer. The shape of these structures implies a combination of mechanical mixing and crack propagation. The grooves on the surface are an indication of plowing. During this process, material deposited aside could also cause a lamellar subsurface structure to emerge.

Very similar results were recorded after the pin slid in one direction only. The main difference was the fact that the grains are morphologically oriented in the sliding direction. Cross-sections perpendicular to the sliding direction revealed vortex formations. This structure is demonstrated in Figure 3.6. Vortexes are reported to appear under surface grooves [48, 49]. The formation of vortexes may be related to pores and cracks inside the shear band in these experiments, where the lamellae begin to form. Loosening in this band increases the eddy motion until particles experiencing pure rotation appear [50]. Vortexes have also been described as the result of vortical "shear+rotation" [51]. From our measurements, we cannot exclude the presence of vortexes in bidirectional sliding.

Figure 3.7 suggests that before the onset of severe plastic deformation and wear on the surface, the sub-surface does not undergo notable changes. The grain structure here was not very different from that of the polished sample, despite the high nominal pressure applied (2.5 MPa).

The main focus of this work is the *in-situ* study of the wear progress on the surface. The generation of wear particles was not observed in the entire field-of-view. The formation of pits mostly occurred on an elevated structure which spreads along the sliding direction. This raised area resembles surfaces with solid lubricating films when blistering occurs [52].

This elevation can be explained by mechanical mixing and material transfer. As the harder body deforms the counter-face, it spreads the deformed volume along the sliding direction, covering part of the surface nearby. Weak adhesion causes separation between the layers. As they pile up randomly on top of each other, deform, and crinkle due to the cyclic load, they give the impression of a continuous lamellar structure. The flatter surface underneath does not conform to the shape of the flakes above, so the separation at this depth is easier to recognize. The oval shape of the spreading area is very similar to the formation of lamellar extrusions as reported in the literature [37]. In this case, the formation of wear particles is not controlled by delamination originating from sub-surface cracks.

Another explanation is that cracks in the nc/ufc layer lead to a local separation, possibly letting lubricant enter the sub-surface. The depth at which sub-surface void nucleation should occur has been studied in literature [53]. In either case, this result is in accordance with the lamellar structure observed in the cross-sections.

The initiation of the elevated structures was not observed. The wear zone appeared from the side of the field-of-view. The height bump appearing at 5065 cycles in Figure 3.9 was either formed directly at this site, or was carried from a neighboring site which cannot be seen during the test. Particles floating to the neighboring area and being captured between the two sliding bodies after every cycle is a possible explanation for the spreading of the area where wear takes place. This, however, is a parameter that cannot be studied in detail in the experiments performed in this work. If this assumption is correct, the mobility of particles on the surface should affect the spreading.

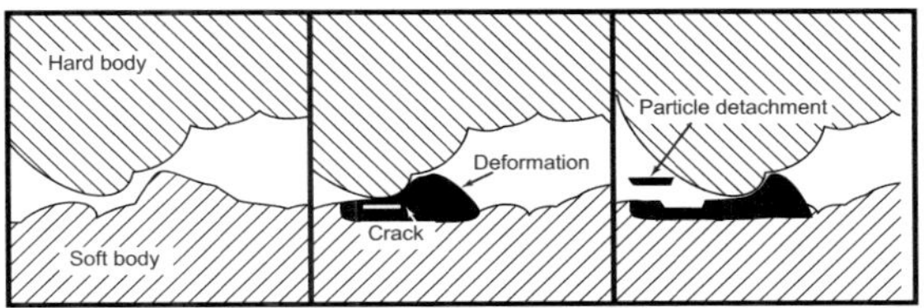

Bild 3.15.: Schematic representation of delamination theory.

In the results section (3.3.4), one could argue that the example of the single asperity moving along the sliding direction, leaving a pit behind, is in good agreement with the delamination theory. The profiles in Figure 3.12 are a result of multiple cycles and appear to agree with the theory of delamination as shown in Figure 3.15 [34, 35]. Nevertheless, this close resemblance does not prove that delamination occurred. The attachment of a particle or transfer of mass onto this site could have formed the asperity observed. If the sub-surface under this asperity was already lamellar, this large volume on a mechanically weak substrate could easily become the reason for instability and hence detachment of the surface layer.

The aforementioned volume loss along with the presence of a thick (60 nm) Cu layer on the pins suggest material transfer from the softer to the harder counter-face. No Fe was detected on the Cu samples, so no transfer is expected to occur in this direction. Besides the observed transfer of Cu to the Fe pin, it is possible that a fraction of the transferred material on the pin returns to the Cu surface.

3.5. Conclusions

In-situ monitoring of the wear behavior of a Cu sliding surface under lubrication was performed to understand how surface deformation, material transfer and wear affect the topography. Upon breach of a thin oxide layer, plastic deformation and particles initiate wear on the surface, leading to a temporary increase of μ until the sample has been run-in. In this process, particle generation occurs in areas of the surface that spread. The mobility of floating wear particles could also play an important role in the propagation of this zone.

Back-up investigations with FIB cross-sections link this work to recent studies of Cu surface/sub-surface microstructure in the steady state. The grain structure becomes finer near the surface, while a nc/ufc structure hosts cracks. Plowing, lamellar extrusions and vortical mechanical mixing may also give rise to crack-like features. The combination of all these effects lead to the final sub-surface structure during sliding. Considering these observations and the *in situ* results, the generation of flake-like particles is not necessarily a consequence of delamination due to sub-surface cracks.

Surfaces loaded with similar conditions have a dependence on the direction of sliding. If the pin slides in one, non-reciprocating direction, the grains appear to be drawn in this direction toward the surface. The reduction of the grain size depends more on wear and plastic deformation in surface material than on the mechanical load itself.

XPS profiles on the counterfaces show material transfer from the softer to the harder body but not vice-versa. Still, deposited soft material on the harder surface could return to the softer body.

4. Linking the Friction Coefficient with Topography during Plowing

4.1. Introduction

According to the theory of Bowden, Moore and Tabor [4], the coefficient of friction consists of two terms: the plowing term and the shear term (Equation 4.1). In their experiments, the aforementioned authors investigated plowing for single pass tests of hard hemispherical spades rubbing on metallic surfaces. It was shown that the flow pressure for plowing p' was not identical with the flow pressure for indentation p. This either indicates different mechanisms that govern the behavior of the contact, or non-identical material properties between the surface and the bulk. In many experiments, the first pass includes material changes near the surfaces, that only occur in this phase. For this reason, the data gathered during the first cycle of an experiment are often neglected. Considering the above, the initial experimental approach of Bowden and Tabor fails to describe plowing flow pressure for advanced cycles. A better examination of plowing after the initial cycle calls for *in-situ* measurements.

$$F_F = A'p' + A_r s \tag{4.1}$$

$A'p'$ is the plowing term and Ap is the shear term.

An approach suggested to measure plowing *in-situ* is to use an indenter to scratch a surface and at the same time monitor the indentation of the probe [54]. The problem that arises in such measurements, is that the elastic recovery cannot be considered [55].

At the nanoscale, atomic force microscopy can be used for *in-situ* studies of plowing [56]. Such experiments are however limited to tribological loads applied at the nanoscale. Therefore, they cannot be used to examine complex lubricated multiple-asperity systems.

During running-in, a variety of complicated processes take place, making it impossible to estimate which part of the measured friction belongs to each of the terms that appear in Equation 4.1. This means, that the processes that are described in Chapter 1 lead to a time dependence of the parameters in Equation 4.1. Again, *in-situ* methods are necessary to observe physical and chemical changes of the third body as cycles progress.

Clarifying the contribution of each source is very useful when planning improvements for the conditioning of a tribosystem. The evolution of shearing and plowing can indicate what parameters or preconditioning measures may assist toward better running-in. Tweaking according to the monitored changes may lead to novel procedures that have not yet been explored. To date, plowing cannot be monitored during running-in and therefore its behavior remains unexplored.

It is however known, that the plowing term strongly depends on the bulk properties of the material and the sub-surface layers, whereas the shear term has a greater dependence on the nature of the intimate junctions between the two metallic surfaces, without, however, being completely independent from the near surface volume. More recently developed models derived from experiments of a diamond spherical tip on nitrided steel suggest that, in this system, the plowing term is independent of the surface condition and the shear term exclusively depends on the surface condition [55].

Plowing in lubricated metallic surfaces is mainly caused by harder asperities of the counter-face and particles entrapped between the two surfaces. The diversity of the geometries observed in these media has triggered research on the topography of wear scars. Based on their research, Komvopoulos et al. built a theoretical model which links geometrical parameters of asperities, particles and plowed zones to the friction coefficient [57]. It was shown, that the sharpness of asperities and particles play a dominant role in plowing and lead to abrasive wear [58].

Unfortunately, to date, there are no methods that allow for an efficient separation of the plowing and shear term of Equation 4.1 when nano- or microplowing occurs in lubricated multi-asperity systems. When attempting to examine such systems, tribologists face the following difficulties:

- Most instruments only measure the friction forces, while wear or surface deformation is only determined with topographical methods after the application of mechanical load.

- On-line wear measurement with RNT has been proven very powerful, but only the amount of detached material can be determined. Plowing involves moving material aside without necessarily leading to wear particle generation.

- Existing *in-situ* optical methods with transparent media monitor the entire wear track, reducing the resolution of the measurement when trying to determine small changes in the area where plowing occurs. Furthermore, these experiments are limited to transparent counterfaces, so the adhesive interaction of various metal couples cannot be investigated.

- Complicated contact geometries make the calculation of plowed volume difficult.

A considerable effort to model plowing and shear friction coefficients during high-temperature ball on disk tests was presented by Wang et al. [59]. In this work, based on existing scratch models and a detailed calculation of the contact interface, a model was proposed to determine plowing and shear friction. Experiments were carried out to confirm the presented model. In these tests, the width of a wear track, formed by a steel ball (6 mm diameter) on an aluminum disk was measured every 10 cycles. Although the contribution of plowing to the friction force was estimated below 1 %, the measured changes of the wear track width were in the order of hundreds of μm. Such plastic deformations cannot be compared to micro- and nanoscopic plowing, which affects the running-in process.

The new approach presented in this chapter takes advantage of a novel tribometer used for high resolution monitoring of plowing while sliding. Furthermore, the possibilities of a simplified approximation for the calculation of plowing and shear terms is put to the test.

4.2. Experimental Setup and Procedure

4.2.1. Tribometer and method

The instrumentation used in this work is capable of monitoring changes on the surface during a sliding test between non transparent metals. Details on the experimental layout can be found in Chapter 2. In a normal experiment with a flat metallic pin sliding against a flat plate, the plowing areas within the wear track spread at random positions. The microscope has a field of view (FOV) of about $80\,\mu m \times 80\,\mu m^2$, which on the one hand provides the necessary lateral resolution for the measurement, but, on the other hand, it does not allow for monitoring of all plowing tracks simultaneously.

The aim of the experiments below is to simplify the tribosystem and calculate p' after monitoring the plowing that occurs during each cycle. In these measurements, it is crucial that other influences affecting the friction coefficient are minimized. The gradual formation of a third body during running-in would entail constant changes in the shear behavior of the junctions. To isolate and study the plowing term, it is imperative, that the friction force is dominated by plowing.

Considering the above, experiments were performed with a ruby sphere sliding in a reciprocating motion against a Cu plate in lubrication. The criteria that lead to the selection of these samples are the following:

1. The selected spheres have a very well-defined geometry that allows for easy calculation of the plowing cross section.

2. There is only one plowing track, so it is not necessary to move the microscope along the width of the contact area to measure smaller plowed areas within.

3. No plastic deformation is expected to occur on the ruby surface, as it is significantly harder than Cu.

4. Ruby is an inert material, so no chemical reactions should occur on the surface. This prevents changes in the nature of the intimate contact between the two bodies.

5. The lubricant prevents the accumulation of debris in front of the slider, which could affect the friction values.

6. Using a sphere leads to higher nominal pressure, ergo higher wear rates, probably too high to establish conditions within the running-in corridor (Figure 1.2). This minimizes the effect that the formation of a third body has on the friction coefficient values.

An additional axis with a screw-drive adjust was fixed to the optical microscope. This allows it to move horizontally and perpendicular to the sliding direction. With this addition, the microscope can be brought to the edge of the wear track. As cycles progress, the track becomes wider, shifting the edge to a new position. Once the edge has reached the end of the FOV, the position of the microscope is readjusted. At certain time intervals the microscope is moved to both edges to measure the total width. This is achieved with a scaled indication of the axis, with a precision of $2\,\mu m$.

Further experiments were conducted with flat steel pins. The purpose of using a multi-asperity system is to investigate if plowing in smaller sections of the apparent contact area also correlates with F_F.

4.2.2. Materials and experiment conditions

All friction/wear experiments were conducted in laboratory conditions: temperature, $T = 25\,°C$, humidity, approx. 45 %. *In-situ* topography images were acquired after every loading cycle with a holographic microscope (DHM® R1000 series, Lyncée Tec SA). As the objective lens remained in immersion during acquisition, a transparent PAO-8 oil with a viscosity of $45.8\,\text{mm}^2\,\text{s}^{-1}$ at $40\,°C$, provided by Fuchs Petrolub AG, was selected as lubricant. The oil circulation was performed by a pump integrated in the radio nuclide technique apparatus (Zyklotron RTM 2000 of Zyklotron AG, Germany). During the experiment, the lubricant's temperature was stabilized at approx. $33\,°C$. The plate samples were made of Cu (purity: 99.98 wt%) and polished with a $1\,\mu m$ diamond particle suspension without further treatment. The average grain size measured on the surface was $1452\,\mu m^2\pm71\,\mu m^2$. The flat pins were made of bearing steel (100Cr6/0.9-1.05 % C and 1.35-1.65 % Cr). The contact area was circular in shape with a diameter ranging from 1.6-2 mm depending on the edge taper. These samples were polished before testing in the same way as the Cu plates. The pin samples had a slight curvature due to the polishing procedure. The measured root mean square roughness (rms) was measured for both sample types. The rms roughness was approx. 200 nm. The roughness was recalculated after subtracting a spherical term in order to eliminate the curvature factor, leading to rms values of approx. 26 nm. The ruby spheres used had a diameter of 1 mm. The roughness (rms) of the samples was about 225 nm and 13 nm after filtering out the curvature.

The sliding experiments were performed in a linear reciprocating motion at various normal loads and speeds, in the range of 2-12 N and 10-25 mm s^{-1}. The normal loads and friction forces shown in the results are recorded at the same area where the sample is examined with the optical microscope. To determine normal and lateral forces at this position, the values are interpolated from the closest measured points in the two sliding

Tabelle 4.1.: Activation parameters for the Cu samples and the reference specimen.

Size of reference specimen	0.781 mg
Specific activity	$a_{spec} = 10.5\,\text{kBq/mg}$ Zn-65
Total activity for Cu sample	$A = 1.8\,\text{MBq}$ Zn-65
Energy windows	50 keV-1400 keV and 430 keV-1400 keV $(E\gamma)$
Linear range	$S_M \geqslant 50\,\mu\text{m}$

directions during one cycle. The initial normal load was adjusted to 2 N for the first 10 cycles of every experiment to avoid deeper damage of the sample due to the high nominal pressure developed before the initial plastic deformation.

4.2.3. Additional experiments

One of the Cu samples was activated at a cyclotron facility (Zyklotron AG, Germany). A protective mask limited the incident protons to a 3 mm thick strip on the specimen, perpendicular to the sliding direction. The area where the track is examined with the optical microscope lyes within the activated region. After irradiation, ^{65}Zn isotopes are expected to replace a small percentage of the ^{63}Cu atoms.

The activation parameters and properties of the sample are gathered in Table 4.1.

The activation and measurement procedure is further described in Figure 4.4.

A Zeiss XB 1540 focused ion beam (FIB) was used to investigate the structure of the samples underneath the worn areas. In order to observe the depth profile, cross- sections along and perpendicular to the wear tracks were milled with a Ga ion gun. No mask was required for the purposes of this analysis, as no thin slices for transmission electron microscopy (TEM) investigations were produced. The images were acquired with a secondary electron detector.

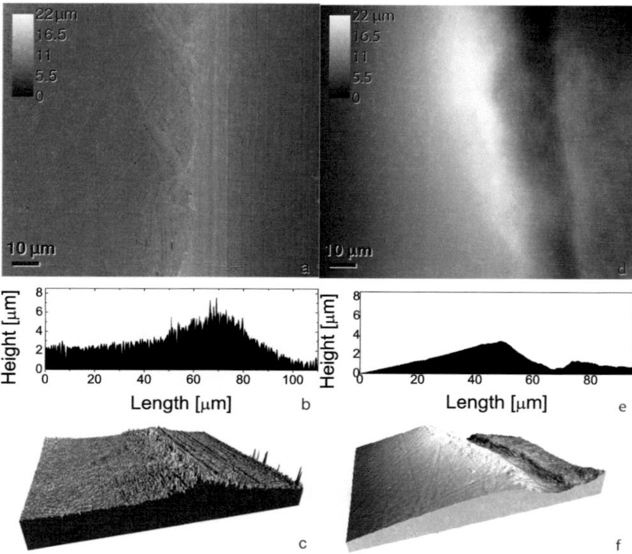

Bild 4.1.: Ridge of a wear track after 50 cycles (a,b and c) and the same ridge after 800 cycles (d, e and f).

4.3. Results

4.3.1. Monitoring wear track edge

For the majority of the results shown in this section, the ruby sphere is used as slider. Plowing causes the formation of a ridge on the side of the wear track. After the first cycle, the microscope has to be brought to the side of the track in order to monitor the position of the ridge. The ridge evolves after multiple passes of the slider. An example after 50 cycles is presented in Figure 4.1 (a,b and c). The height of the ridge reaches 3-5 µm at this stage. Linear structures appear directly next to the wear track. The direction of groups of such lines gathered close to each other appears to be the same and it may change after several cycles of loading. After the fiftieth cycle it is still clear where the end of the wear track is located. After several cycles, the sides of the wear track become vague because of the geometry of the deformed Cu. A thick layer of plowed material covers the intact surface right next to the ridge. This is shown in Figure 4.1 (d, e and f), for the same experiment, after 800 cycles.

Single images can be used to determine the position and the topography of the ridge, but they contain no information concerning its evolution. A series of images after every cycle, better yet a video, could give a more complete idea of the changes that occur.

4.3.2. Profile extraction and mapping

To quantify the shift of the wear track side, profile lines, perpendicular to the sliding direction are extracted at the same coordinates within the FOV for successive images. These lines are placed next to each other to form a profile map. The mapping process is shown in Figure 4.2. The original image on the top left consists of two main areas. The lower, brighter area is the part of the surface that has not yet come to contact with the sphere. The gradient that appears in the upper area corresponds to the worn area that conforms to the shape of the slider. The edge is the thin area that separates the intact from the plowed surface.

As mentioned above, the wear track grows out of the FOV, so at certain time intervals it is necessary to move the microscope. This causes a discontinuity in the profile map. To overcome this problem, the segments of acquired maps are stitched to produce a continuous image with respect to the position of the plowing ridge. The stitching process is shown in Figure 4.3.

A further issue that makes it difficult to distinguish the location of the ridge in the profile maps is a slight vertical shift of the topography images that may occur during reconstruction, especially when the microscope has been readjusted. This problem is evident between 300 and 400 cycles of Figures 4.2 and 4.3. To avoid errors related to this fluctuation, the intact

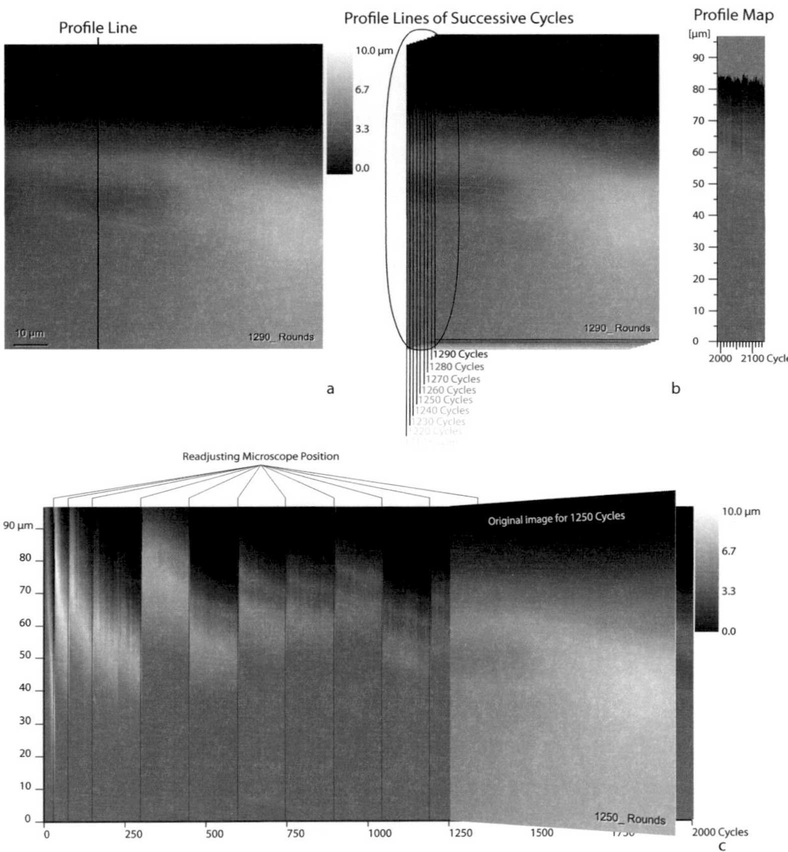

Bild 4.2.: Gathering profiles from the same coordinates of every cycle (a) and placing them next to each other forming a profile map (b). Indicating positions where the microscope moved to acquire the total width of the wear track (c).

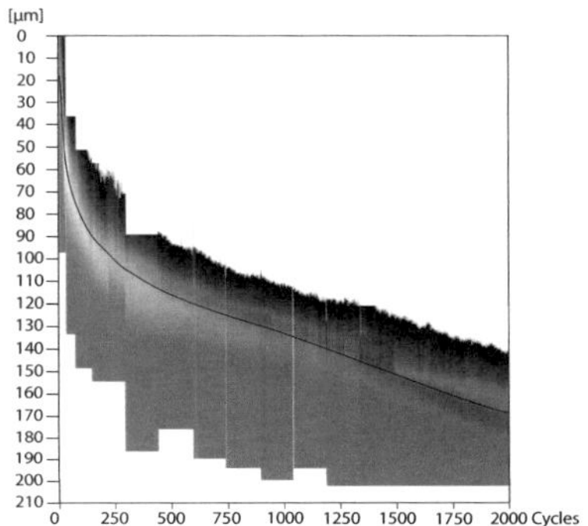

Bild 4.3.: Stitching together the segments between microscope positions and extracting the widening data.

area is used to bring all profiles to the same height level. In this process, the last 50 pixels of every profile are averaged. According to the resulting value, the entire profile is adjusted, so that the polished area is brought to a height level of $0\,\mu m$.

Once the stitching and height corrections are complete, a graphical representation of the ridge position can be constructed. The position of the ridge in these images does not indicate the width of the wear track, but the change of the position shows how fast the wear track widens. Further information on how the widening profile extraction is performed is shown in section 4.4.2. As shown in the example of Figure 4.3, rapid widening occurs for early cycles, the widening then stabilizes into a rather straight line and in many cases ceases entirely.

4.3.3. Experiments with radioactive nuclides

Wear measurements are performed with the RNT apparatus to compare the plowed volume to material that entirely detaches from the Cu surface. The instrument's software automatically calculates radioactive mass in the lubricant, according to the measured intensity and total volume of fluid. The

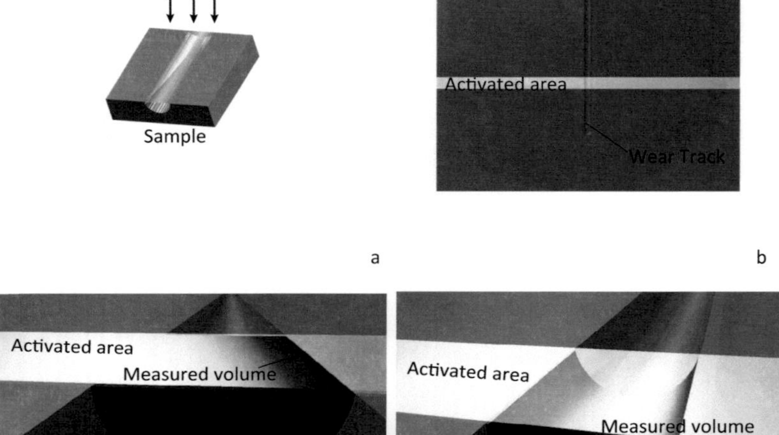

Bild 4.4.: RNT procedure. The sample is activated with a proton beam produced by a cyclotron apparatus (a). Exposed area under mask (b). Theoretically measured volumes if the entire deformed volume produces wear particles entering the oil circulation (c and d).

material loss of Cu is calculated from the specific activity of the sample. Converting the mass to volume and dividing it by the width of the activated strip, as shown in Figure 4.4, gives the cross section area of wear. This area is compared to the cross section area of plowing, as calculated from the indication of the scaled screw-drive adjust of the optical microscope (r_r values). An example of the first RNT experiments conducted is shown in Figure 4.5. The greatest part of the widening in the beginning is attributed to wear and later only a small fraction of A' is detected with the RNT ap-

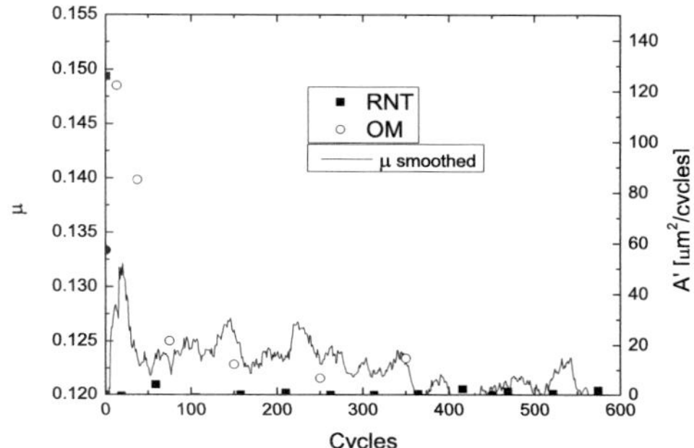

Bild 4.5.: Comparison of plowing cross section and wear. The square points show the cross section area that was calculated from the amount of wear particles that enter the oil circulation, as measured by RNT. The circular points are the plowing cross section area values. The measured μ is also plotted, to investigate possible correlation. The experiment was performed at a normal force of 4 N and a speed of 20 mm/s.

paratus, as the plastically deformed volume only shifts to the sides of the track without leading to wear particle generation.

The RNT results do not exhibit good repeatability. Later experiments show the opposite effect. Namely, no radioactive material is detected for early cycles, whereas some wear appears after about 500 cycles. The reasons for this discrepancy are being investigated.

In any case, no correlation between friction coefficient and wear volume, or wear cross section area are observed in any of the experiments performed.

4.3.4. Sub-surface investigation

The surfaces are further examined with an SEM. Some of the wear tracks exhibit a wavy pattern on the surface. The period of this pattern is approximately 500 nm. Figure 4.6 shows a comparison of a wear track that has

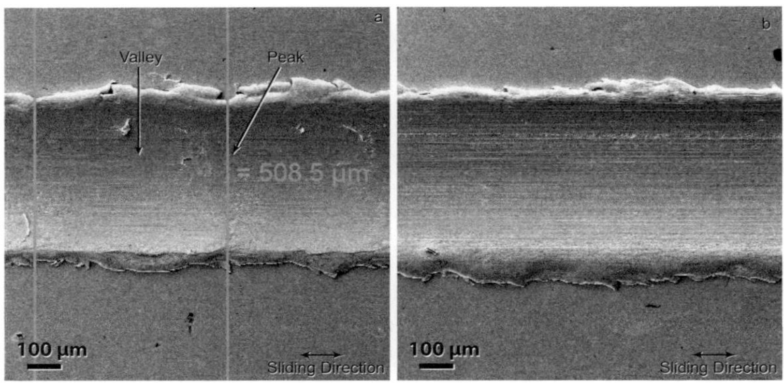

Bild 4.6.: Comparison of wear tracks with (a) and without (b) height fluctuations along the sliding direction. The images were acquired with SEM.

height fluctuations along the sliding direction and another one that does not.

FIB cross sections of the sample that does not exhibit waviness reveals no nano or ultrafine crystalline structure directly under the surface. The grains are evidently finer than in the bulk of the material down to sizes below $1\,\mu$m. The sub-surface of samples with the aforementioned waviness are less homogeneous. In the lower areas between two successive peaks, the structure under the surface is very similar to that observed before. When examining cross sections under the peaks, a fine crystalline and lamellar formation is seen. Examples of these structures are seen in Figure 4.7. The thickness of the lamellar structure in this case is approximately $1\,\mu$m. It can be assumed, that material is accumulated in a loose form in front of the sphere. Once it reaches a critical amount, the sphere skips it and slides on. FIB investigations have not been repeated for enough samples to perform statistical analysis on the formation and properties of these structures.

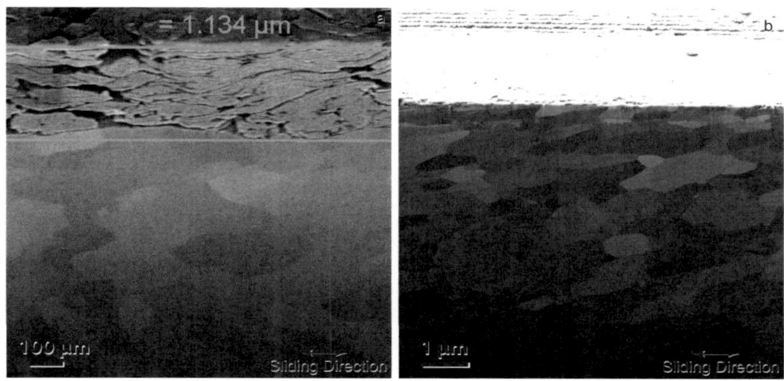

Bild 4.7.: Comparison of subsufrace structures beneath the peaks (a) and valleys (b) of the wavy pattern. The sections were performed with FIB and the images were acquired with SEM.

4.3.5. Experiments with flat pins

When conducting experiments with flat pins, the FOV is randomly placed within the apparent contact area. Although the pins are flattened and polished, a slight curvature is still present. Therefore, a wider plowed area appeared near the center of the contact between the pin and the flat sample. Smaller plowing tracks appear after several cycles around the central plowing track. A representative experiment of contact with a more profound plowing track in the middle is selected to record the response of the friction coefficient to plowing. The FOV is placed at the ridge of one of the main plowed area. The correlation between μ and $\dot{r}$ is shown in Figure 4.8. When comparing the two curves, there are cycle ranges with an obvious mismatch, but at many other ranges, there is a very strong correlation between the two. Even finer details in the behavior of the two lines are common in both μ and $\dot{r}$. Such an example are the peaks at 1250 and 2500, which seem to match. The range between 500 and 1100, as well as between 2000 and 2400, on the other hand, show no evident correlation.

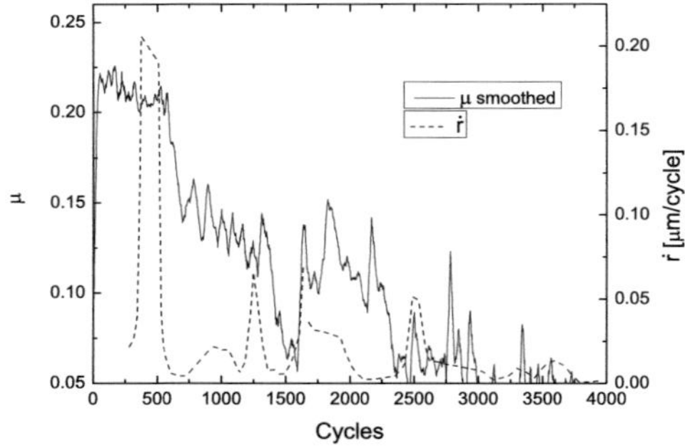

Bild 4.8.: Comparison of the behavior of μ and $\dot{r}$ for a plowing track within the contact area of a flat pin. The values of μ have been smoothed with an averaging algorithm to remove noise. The experiment was performed at a nominal pressure of 1.33 MPa and a speed of 20 mm/s.

4.4. Discussion

4.4.1. Wear track edge

When performing experiments with ruby spheres, the ridge on the side of the wear track can be monitored easily for a limited range of cycles. After multiple passes, debris is deposited around the contact area. Although the oil carries away a great part of these particles, some still adhere around the wear track, making it hard to distinguish between worn and intact area. This is also visible in the SEM images, in Figure 4.6, although the samples were cleaned with hexane and aceton before the SEM acquisition.

4.4.2. Correlating widening and friction

The profile extraction procedure has been described in section 4.3.2. Once all profile segments are stitched together and the zero level of the images is set, axes are drawn around the image to quantify the data. The position of the ridge is marked by placing data points over the map to extract a graphical representation of the ridge shift as the cycles progress. These data points are placed by hand. The first measurement of the position is performed after the first cycle of the slider. Therefore, the radius r is defined as the position of the ridge with respect to its position after the first cycle. The real radius value in this case would be $r_r = r + R_i$, where R_i is the initial width of the track after the first cycle. For simplicity, the symbol r will be used as widening on one side of the wear track. Given that the slider has a perfectly spherical shape, it is assumed that the same widening should be occurring on the opposite side of the track. Therefore, the total widening should be given by multiplying this value by 2.

The evolution of r can be monitored as shown in Figure 4.3. When comparing r with the μ values against time/cycles, there appears to be some correlation, as both seem to change dramatically in the beginning and start stabilizing after several cycles. Such an example is shown in Figure 4.9. Plowing will continue to occur, causing further widening, so when the μ values become rather constant, there is no further connection between μ and r. The two graphs do not seem to correlate directly. There is a better correlation when using the widening rate $\dot{r}$ instead of the widening itself. This value is calculated by differenciating the r plots against the number of cycles. In this case, there is a much closer correlation between $\dot{r}$ and μ. Still, below 100 cycles there appears to be a misfit. Between 100 and 200 cycles, the behavior $\dot{r}$ is practically identical to that of μ.

A further example is shown in Figure 4.11. Here, it appears that changes in the μ values are recognized earlier than in the $\dot{r}$ curve. This is seen at about 210 cycles and after 500 cycles. Furthermore, features of the μ curve between 330 and 430 cycles are not reflected in the behavior of $\dot{r}$ at all. Again, there appears to be no connection between the two curves below 100 cycles. For the rest of the cycle range, the two curves seem to have a very similar behavior.

Graphs like the ones shown in Figures 4.10 and 4.11 were used to form a mathematical expression that correlates the friction coefficient with the widening rate. In order to bring the two curves at the same level until the

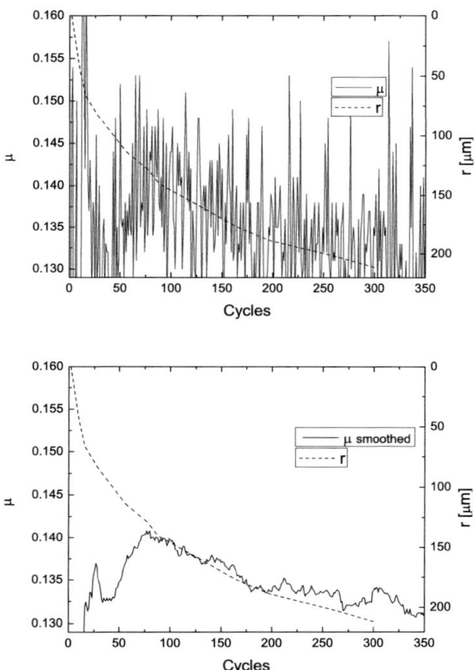

Bild 4.9.: Comparison of the behavior of μ and r. In the bottom image, the values of μ have been smoothed with an averaging algorithm to remove noise (30 point AAv average smoothing with Originlab Origin). The experiment was performed at a normal force of 8 N and a speed of 20 mm/s.

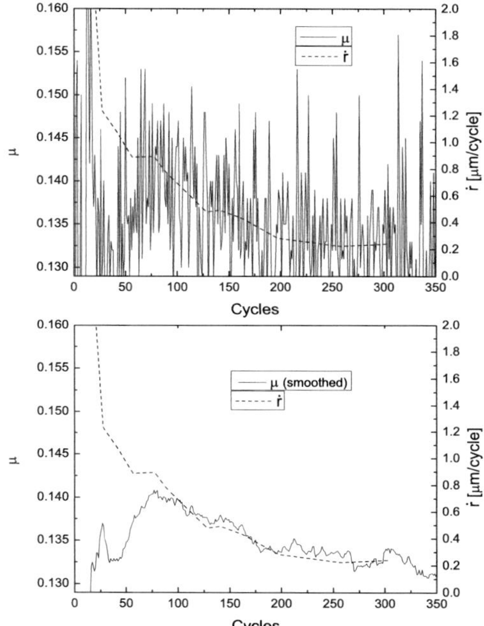

Bild 4.10.: Comparison of the behavior of μ and $\dot{r}$. In the bottom image, the values of μ have been smoothed with an averaging algorithm to remove noise (30 point AAv average smoothing with Originlab Origin). The experiment was performed at a normal force of 8 N and a speed of 20 mm/s.

common behavior of the two can be noticed, the range of the axes is furtger adjusted. Figure 4.12 gives a general example, where (μ_{min}, μ_{max}) is the range of the friction coefficient and $(\dot{r}_{min}, \dot{r}_{max})$ is that of the widening rate. If the two curves have practically identical paths for these ranges, then the mathematical expression that connects $\dot{r}$ with μ is

$$\mu = a + b\dot{r}, \tag{4.2}$$

where

$$a = \mu_{min} - \dot{r}_{min}$$

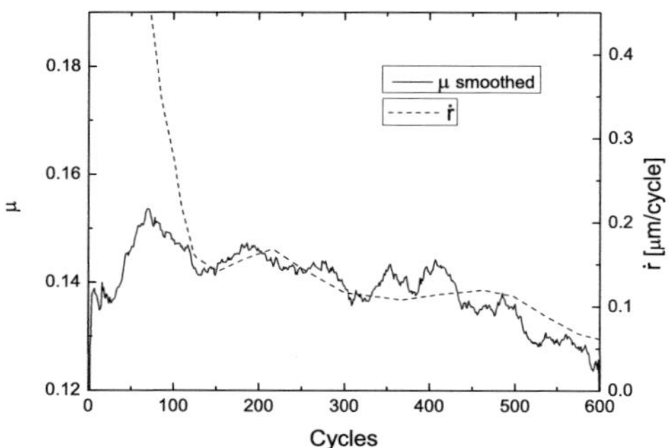

Bild 4.11.: Comparison of the behavior of μ and $\dot{r}$ for a wider range of cycles. The values of μ have been smoothed with an averaging algorithm to remove noise (30 point AAv average smoothing with Originlab Origin). In this case, the experiment was performed at a normal force of 4 N and a speed of 15 mm/s.

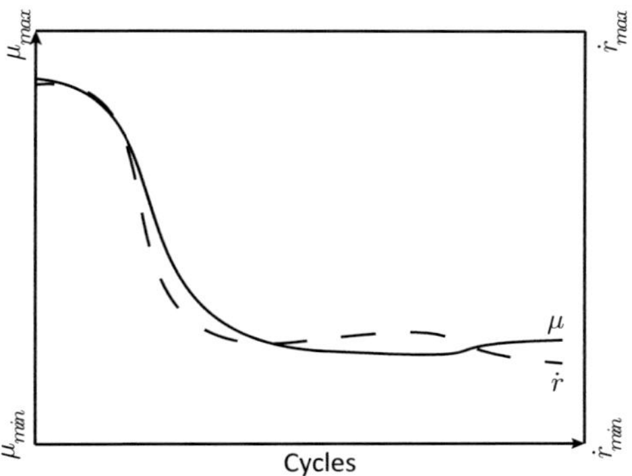

Bild 4.12.: Example of range values extraction for the calculation of the correlation parameters between μ and $\dot{r}$.

and

$$b = \frac{\mu_{max} - \mu_{min}}{\dot{r}_{max} - \dot{r}_{min}}.$$

The constant b also contains the factor 2 which corresponds to the widening recorded on the other side of the wear track, assuming that deformation is equal on both sides.

To determine the dependence of these parameters on the sliding speed and normal load, the experiment was repeated for all combinations of F_N of 4, 6 and 8 N and v of 10, 15 and 20 mm/s. No particular trend is observed, there are only random deviations from the average values.

4.4.3. Plowing cross section area

To confirm that the above approximation is consistent with the theory of Bowden, Moore and Tabor, the behavior of μ is also directly compared to that of A'. To calculate A', r_r is necessary. The total width can in principle be calculated as shown above ($r_r = r + R_i$) if R_i is known. In this approach it is assumed, that the widening on either side of the wear track is always equal. To ensure that no differences between the two sides of the wear track could affect the results, the screw-drive adjust indication is used to measure the total width of the track after certain time intervals.

In a multiple pass experiment, the geometry has to be redefined. Figure 4.13 shows a comparison between the geometry of the first pass and that of an advanced cycle state. Both the real contact area A_r and the plowing cross section A' are significantly larger for the first pass. The calculation of A' has been studied thoroughly for various geometries of sliders [60, 57]. Calculations of multiple pass cross sections have already been performed in the literature [59]. For the purposes of this work, simplified geometric calculations that do not consider elastic recovery are applied.

To determine the geometry of A' in the case of multiple passes, sphere and counter-face are reduced to a two dimensional representation. If the vertical position of the sphere is shifted downward, toward the counter-face,

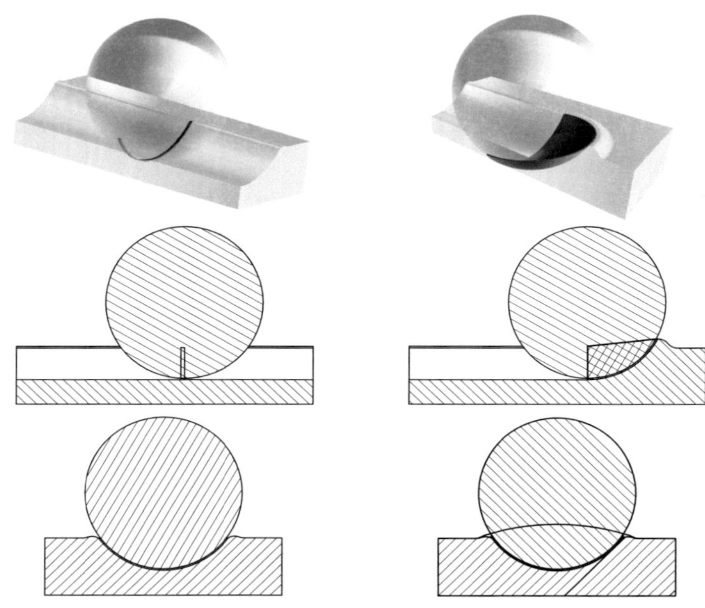

Bild 4.13.: Comparison of the geometry of the real contact area and the plowing cross section for single and multiple pass experiments.

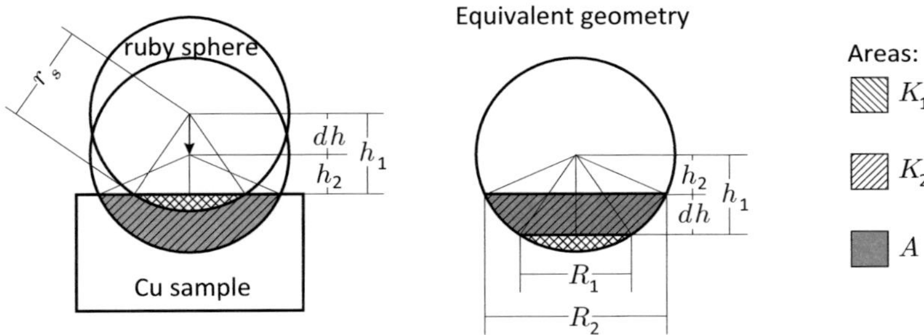

Bild 4.14.: Calculation of the plowing cross section A' as a difference between circle segments K_1 and K_2 for an indentation of dh.

by dh, and the distance of the center of the circle from the counter-face is initially h_1 and after a cycle h_2, then $dh = |h_2 - h_1|$. This can also be described as an "indentation" of the sphere. Figure 4.14 shows the geometry of gradual indentation dh of a sphere with radius r_s in a flat surface. The area that needs to be calculated is A' and it can easily be found by the difference between the circle section areas K_1 and K_2: $A' = |K_2 - K_1|$. The equations used for the calculation of circle sections and arcs are given below:

$$L = r_s \arccos\left(\frac{c}{2r_s}\right), \tag{4.3}$$

$$K = \frac{1}{2}\left[r_s^2 \arccos\left(\frac{R}{2r_s}\right) - R\sqrt{r_s^2 - \left(\frac{R}{2}\right)^2}\right]. \tag{4.4}$$

These equations are based on geometrical calculations according to Figure 4.14.

If n cycles have been performed between two successive width measurements, the calculation of the average plowing cross section area is given by the following equation:

$$A' = \frac{|K_2 - K_1|}{n}. \tag{4.5}$$

Figure 4.15 shows a comparison of the evolution of μ and A' for the same experiment presented in Figure 4.10. The precision of this measurement is significantly lower than monitoring the evolution of the ridge after every cycle as shown above. This makes the correlation between the two curves very difficult. It is seen, that the values are scattered around the μ curve, especially in the first half of the graph. Nevertheless, the range of the values seems to remain at consequent levels for μ and A'. Although the cross section areas remain at particularly high levels in the beginning of the experiment, they do not come off-scale as in the case of $\dot{r}$, as shown in Figure 4.10.

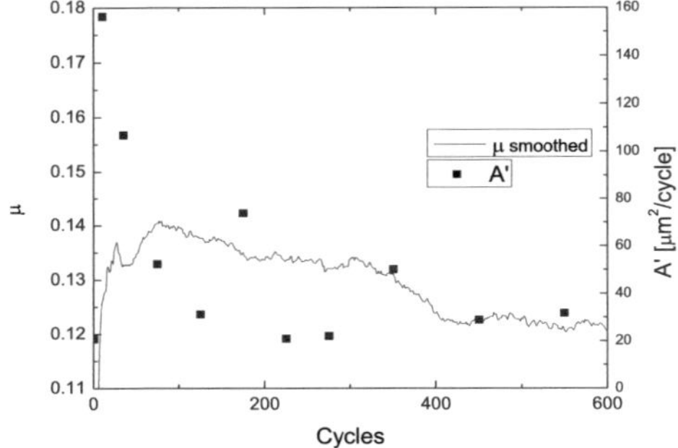

Bild 4.15.: Comparison of the behavior of μ and A'. The values of μ have been smoothed with an averaging algorithm to remove noise (30 point AAv average smoothing with Originlab Origin). The experiment was performed at a normal force of 8 N and a speed of 20 mm/s.

4.4.4. Approximating the plowing term

Despite these difficulties, it appears that the connection between μ and $\dot{r}$ is evident in Figures 4.10 and 4.11. Equation 4.2 initially seems to contradict the theory of Bowden Moore and Tabor (see equations 1.1, 1.2 and 1.3). This is however not the case. The geometry of the cross section area of plowed volumes is a very thin strip along the arc of the spherical slider. If the indentation dh is significantly smaller than the total width of the wear track, then A' can be simplified to a long parallelogram with two triangles on either side, as shown in Figure 4.16. These triangles only occupy a very small part of the total area. Furthermore, the widening during every cycle is significantly smaller than the total width of the track. According to these assumptions, the factor that connects $\dot{r}$ with A' is dh.

The relation between r and dh is

$$r = \tan(\theta)\,dh.$$

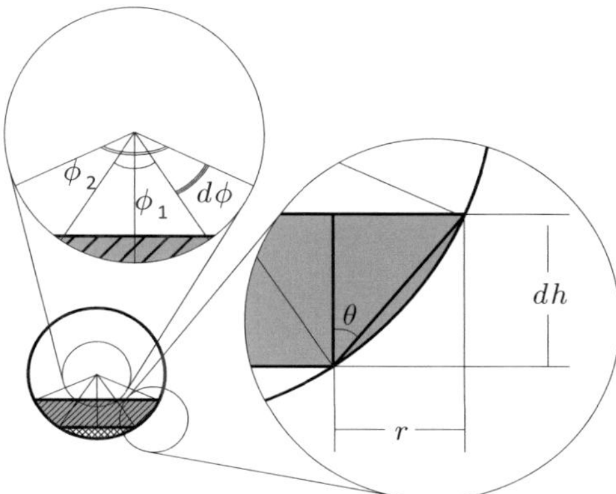

Bild 4.16.: Example of range values extraction for the calculation of the correlation
parameters between μ and $\dot{r}$.

If the difference between the arc angles $d\phi$ of two successive cycles is very
small, then, for a limited cycle range, the connection between $\dot{r}$ and dh is
practically linear.

In most tribo-systems, running-in leads to the formation of a third body
as described in Chapter 1. With the exception of the peaks observed in
wear tracks with a wavy structure, no such third body was observed for
experiments conducted with ruby spheres. Plowing and wear levels are so
severe, that proper running-in is not occurring. This means that in the curve
shown in Figure 1.2, the conditions for these experiments are on the left
side of the running-in corridor. According to Coulomb's law of friction, the
apparent contact area does not affect the real contact area. The normal load
for all experiments is held constant after the first ten cycles. Considering
all the above, no significant changes affect the shear term of the friction
coefficient. Therefore, μ is controlled by plowing in these experiments,
and $A_r s$ is considered constant.

This means that the two terms of Equation 4.1 can also be found in Equation 4.2, approximated as follows:

$$\frac{A_r s}{F_N} = a,\tag{4.6}$$

$$\frac{A' p'}{F_N} = b\dot{r}.\tag{4.7}$$

4.4.5. Approximation critera

This approximation only holds if the following conditions are fulfilled:

1. The running-in corridor conditions are not fulfilled.

2. The indentation occurring between two cycles is significantly smaller than the total wear track width, $dh \ll r_r$.

3. The approximation is applied for a short range of cycles.

4. No significant changes occur in the geometry of the slider.

5. The slider's surface is not affected chemically while sliding.

According to the above criteria, it is possible to explain why the approximation fails for cycle ranges below 100. During the early cycles, dh is comparable to r_r. Furthermore, F_N is so high, that the surface damage is not only caused via plowing. Lines appear on the side of the plowing track, that seem to be following the persistent slip bands (PSB) of the grains. This means that deformation is caused in a much larger area around the contact area. This energy loss is also expressed as an increase of resistance against sliding, hence higher μ.

It is possible, that at very high cycle numbers, the normal load is distributed on a much larger area, possibly bringing the system into the running-in corridor. In addition, if this approximation is applied to a very large number of cycles, it is possible that the values of a and b are not consistent throughout this range. The reason for this is that as the contact area in the shape of an arc, as shown in Figure 4.14, grows significantly, but the radius of the sphere remains the same. This geometrical change may affect b. In addition, the geometrical change may also affect the angles at which the normal load is distributed, as the sides of the area near the ridges are no longer near a horizontal orientation, so a may be affected as well.

Severe surface damage in the beginning of the experiments leading to the detachment of larger volumes of Cu would explain the behavior of wear shown in Figure 4.5. Due to lacking repeatability of these results, no conclusions can be drawn. It is possible, that the particles generated during early cycles are sometimes too large to enter the lubricant circulation. If large particles are deposited and immobilized near the wear track, or if their flow is hindered at the filter of the RNT apparatus, then they cannot be detected. Further experiments need to be conducted to address this issue.

For the results used in the above analysis, it is assumed that the shear term of Equation 4.1 is practically constant, as no running-in occurs. The cross sections performed with FIB showed that some samples exhibited a waviness. The peaks of these structures had a lamellar subsurface, which is a strong indication of local running-in. Further investigations are necessary to determine if these formations affect the results used for the correlation between the curves of μ and those of $\dot{r}$ or A'.

4.4.6. Estimation of plowing term and flow pressure

Equation 4.7 can be used to calculate the plowing term. This can either be calculated by measuring r when monitoring the wear track ridge (method A), or by measuring the total width with the indicated axis of the microscope (method B).

When comparing the methods presented above:

- Method A has a significantly higher time resolution than B.

- Method A has a better spatial resolution than B.

- Method B determines the exact total width, r_r, regardless of unsimilar widening on either side of the wear track.

- Method A requires an initial measurement of r_r.

In a conventional experiment, the microscope remains at a constant position. As mentioned above, this is not possible throughout the experiment, as the wear track grows out of the FOV and the microscope's position has to be adjusted regularly. Every such adjustment is a good opportunity to perform a total width measurement as a reference.

An ideal experiment is the combination of methods A and B. The initial width after the first cycle is measured with method B. The main calculations are performed with the $\dot{r}$ values from method A. Method B is used to ensure, that the values acquired are reliable, meaning that widening is equal on both sides.

A combination of these methods can also be used to calculate the plowing cross section more precisely. As shown above, the total widening is $2r$. With an offset of $r_i = 60\,\mu\mathrm{m}$, the total width values r_r measured with method B are practically the same as the shifted $2r$ values measured from method A. This is shown in Figure 4.17. As the resolution of method A is higher, the A' values are now calculated with the $r_i + 2r$ curve. The result is again compared to the smoothed curve of μ in Figure 4.18. Working in the same way as in Figure 4.12, the factor b' should now come from the following equations:

$$\mu = a + b'A' , \tag{4.8}$$

where

$$b' = \frac{\mu_{max} - \mu_{min}}{A'_{max} - A'_{min}} = \frac{0.155 - 0.12}{170 \cdot 10^{-12}\,\mathrm{m}^2} \simeq 205 \cdot 10^6\,\mathrm{m}^{-2}$$

and

$$\frac{A'p'}{F_N} = b'A' , \tag{4.9}$$

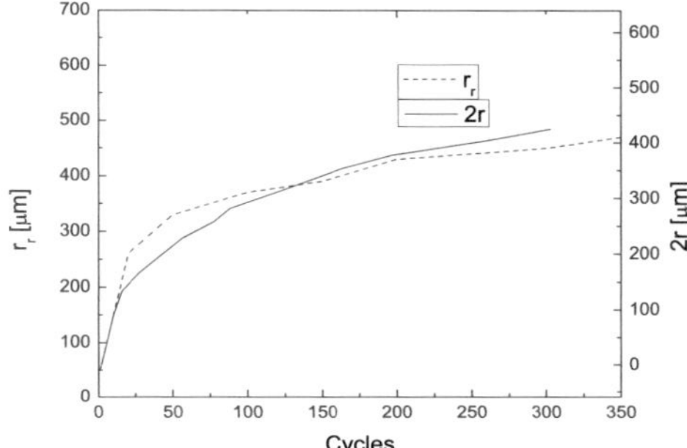

Bild 4.17.: Comparison of the behavior of r_r and $2r$. With an offset of $60\,\mu$m . The experiment was performed at a normal force of 8 N and a speed of 20 mm/s.

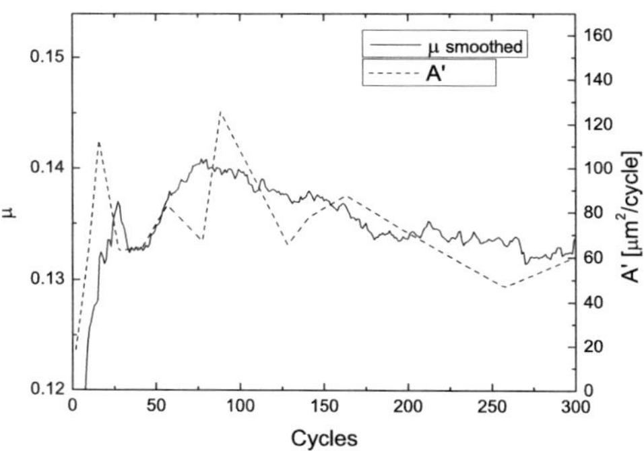

Bild 4.18.: Comparison of the behavior of μ and A', calculated by a combination of methods A and B. The curve of μ has been smoothed. The experiment was performed at a normal force of 8 N and a speed of 20 mm/s.

so

$$b' = \frac{p'}{F_N} \;\Rightarrow\; p' = b'F_N = 205 \cdot 10^6\,\mathrm{m}^{-2} \cdot 8\mathrm{N} = 1.6\,\mathrm{GPa}\,.$$

This results in a flow pressure of $p' = 1.6\,$GPa for the Cu sample used in these experiments.

4.4.7. Multiple-asperity systems

The data acquired in this study needs to be related to a realistic tribosystem. In such cases, the approximation used in Equation 4.2 can be used for multi-asperity systems. In contact mechanics, multi-asperity systems can be dealt with as surfaces composed of hemispherically tipped asperities of uniform radius. This concept was introduced in the model of Greenwood and Williamson. Accepting this simplification, the plowing term depends on the collective contribution of smaller asperities in contact. Figure 4.8 proves, that the behavior of a single plowing track can directly affect the friction coefficient according to the above principles.

The instrument used for this study can monitor the topography within the apparent contact area *in-situ*. If the surface that is being analyzed is located within a spreading plowing track, then the widening rate $\dot{r}$ can give a good indication of the plowing contribution in this area. The widening rate of smaller plowing tracks also gives an idea of the distribution of load throughout the contact area.

Future work could involve building a system that can monitor the widening of all smaller tracks while maintaining the high resolution of the methods presented in this study. For relatively homogeneous wear tracks, one or more representative positions can be used to estimate the amount of plowing that takes place on the entire contact area. Monitoring significantly smaller plowing tracks can be particularly challenging with conventional optical methods; however, succeeding in developing such a technique would enable tribologists to determine A' for multiple-asperity systems. As p' for sliding against a soft body can be fully determined and it mainly depends on the near surface volume of this material, the changes of the shearing term that occur during running-in can be isolated. The last missing piece of the puzzle to fully determine the evolution of τ are real contact area estimations A_r. Admirable work on A_r, conducted in various research groups around the words, is currently in progress. It has to be noted, that deviations from the calculated p' may arise due to geometric and chemical variations on the surface. The criteria for the approximation presented in section 4.4.5 need to be reconsidered, when generalizing the simplified model presented above to multiple-asperity systems. Finally, the results gathered in this work can also be compared to more complex plowing models found in the literature.

4.5. Conclusions

In this chapter, reciprocating linear sliding of a hard, spherical, and inert body against Cu was used as a technique to experimentally separate the plowing and shear terms of the Bowden Moor and Tabor equation. Two main methods were explained:

1. Correlation of the plowing track widening rate $\dot{r}$ with the friction coefficient. The widening r is extracted from profile maps of successive optical image acquisitions at the ridge of the plowing track. An approximation for a limited cycle range can be used for a fast separation of the two terms.

2. Correlation of the plowing cross section are A' with the friction coefficient. The area A' is calculated after measuring the total width of the wear track with an indicated axis attached to the microscope.

The first method exhibits a spatial resolution equal to the resolution of the optical image output of the microscope and a time resolution of 1 cycle. The second method has a lower spatial and time resolution, but it provides full plowing track width r_r information.

A combination of the two methods can be used for a precise calculation of plowing flow pressure p'.

In a multi-asperity system, the approximation used in the first method can provide a fast indication of plowing contribution of smaller regions of the apparent contact area. Future work on real contact area and simultaneous measurement of all plowing tracks growing within the apparent contact area can lead to *in-situ* monitoring of τ during complex running-in processes.

These observations and experimental methods help simplify the complex behavior of tribological systems and show great potential for future development in the field of *in-situ* experimentation in tribology.

5. Summary

In order to shed some light onto the dynamics of lubricated sliding metallic surfaces, a novel tribometer was constructed. The instrument combines a high precision positioning stage and an advanced force sensor with an immersion holographic microscope. The concept behind the novel tribometer combines state-of-the-art methods for the analysis of the surface topography on the micro- and nano-scale with the online measurement of wear. At the same time, it allows for frictional and lateral force detection. Information on the topography of one of the two surfaces is gathered *in-situ* with a 3D holography microscope at a maximum frequency of 15 fps and higher resolution images can be provided at defined time intervals by an atomic force microscope (AFM). The wear measurement is conducted on-line by means of radio nuclide technique (RNT). The quantitative measurement of the lateral and frictional forces is conducted with a custom-built 3D force sensor. The surfaces can be lubricated with an optically transparent oil or water. Tests performed with iron and steel samples demonstrate the stability and versatility of the tribometer.

The first experiments were performed with steel pins against copper plates in poly alpha olefin, sliding in a reciprocal motion, as well as in a uni-directional manner. A general trend that we noticed at all experiments was that initially the friction coefficient remained at extremely low levels (about 0.05), then increased as wear particles formed and caused severe plastic deformation on the copper surface and finally decreased again to values slightly higher than the initial ones. Contrast to previous reports and theories concerning running-in, we did not notice a flattening of the surface during running-in.

The wear particle generation was monitored on-line, with a particular emphasis on flake-like particles formed while sliding. The relevant motion of the slider leads to an elevated area, which spreads during sliding. In addition to the acquired 3D images of the surface, focused ion beam (FIB) cross sections and X-Ray photoelectron spectroscopy investigations were used to better understand the near surface structure. The subsurface structure suggests a mechanical mixing and material transfer mechanism, which leads to an unstable lamellar formation near the surface. Still, delamination cannot be excluded as a possible mechanism in the examined system.

Interestingly, the aforementioned increase of μ was accompanied by the widening of wear stripes parallel to the sliding direction throughout the wear track. Given that these stripes grew in different regions of the contact area, it was hard to correlate their evolution with the friction coefficient.

In order to better understand this effect, the experiment was further simplified by replacing the steel pin with a ruby sphere (1 mm diameter). This sample guarantees that plastic deformation only occurs on the Cu sample, there is only 1 center of the contact area and that the widening of the wear track occurs equally fast on both sides. The microscope was placed at the side of the growing wear track in order to observe the widening of the track after every cycle.

The evolution of the friction coefficient was plotted along with the widening rate. It appears, that when removing a substrate-constant value from μ, it is proportional to this rate in specific cycle ranges. The rate of this widening is connected to plastic deformation, which accounts for the increase of the friction coefficient. It was shown, that this trend can be used as an approximation to separate the plowing from the shear term as defined in the theory of Bowden, Moore and Tabor.

As the contact area in the ruby sphere experiments is initially significantly smaller than that of the steel experiments, it is clear, that the two experiments are performed in different tribological regimes. In the beginning, the surfaces of steel and copper are dissimilar. Therefore, until completion of the running-in phase, the nominal pressure of areas within the apparent contact area are increased. This leads to plastic deformation, which widens these areas until they cover the entire wear track. With the current features of the instrument used, only one of these plowing tracks can be monitored. Nevertheless, a similar connection between widening and μ was observed.

This work forms an innovative approach to *in-situ* tribology of lubricated sliding surfaces. The first results contribute to a better understanding of complex tribosystems. The tribometer displays great potential for improvement. This makes the method developed very promising for further investigations with a wide variety of sliding surfaces. These investigations may elucidate the processes occurring during the complex process of running-in.

Bibliography

[1] D. Dowson. *History of tribology*. Professional Engineering Pub, second edition edition, 1998.

[2] H.P. Jost. Lubrication (tribology) education and research ('jost report'). Technical report, Department of Education and Science, HMSO, 1966.

[3] C. Evans. Strategy for energy conservation through tribology. *Energy Policy*, 6(2):170–171, June 1978.

[4] F.P. Bowden, A.J.W. Moore, and D. Tabor. The ploughing and adhesion of sliding metals. *J. Appl. Phys.*, 14:90 – 91, 1943.

[5] L. Prandtl. Ein gedankmodell zur kinetischen theorie der festen körper. *J. Appl. Math. Mech.*, 8:156–171, 1928.

[6] U. Landman, W. D. Luedtke, and E. M. Ringer. *Molecular Dynamics Simulations of Adhesive Contact Formation and Friction*, pages 463–508. Kluwer, Dordrecht, 1992.

[7] H. Hertz. Über die Berührung fester elastischer Körper. *J. ReineAngew. Math.*, 92:156–171, 1881.

[8] K.L. Johnson, K. Kendall, and A.D. Roberts. Surface energy and the contact of elastic solids. *Proc. R. Soc. Lon. Ser. A.*, 324(301), 1971.

[9] B. V. Derjaguin, V. M. Muller, and Y. P. Toporov. Effect of contact deformations on the adhesion of particles. *J. Colloid. Interface Sci.*, 53:314–325, 1975.

[10] B. V. Derjaguin, V. M. Muller, and Y. P. Toporov. On two methods of calculation of the force of sticking of an elastic sphere to a rigid plane. *Col. Surf.*, 7:251–259, 1983.

[11] D. Maugis. Adhesion of spheres: The jkr-dmt transition using a dugdale model. *J. Colloid. Interface Sci.*, 150:243–269, 1992.

[12] J.A. Greenwood and J.B.P. Williamson. Contact of nominally flat surfaces. *Proc. Roy. Soc., London, Series A*, 295:300–319, 1966.

[13] J. Pullen and J.B.P. Williamson. On the plastic contact of rough surfaces. *Proc. Roy. Soc., London, Series A*, 327:159–173., 1972.

[14] B.N.J. Persson. Relation between interfacial separation and load: A general theory of contact mechanics. *Phys. Rev. Lett.*, 99(125502), 2007.

[15] J.F. Archard and W. Hirst. The wear of materials under unlubricated conditions. *Proc. Roy. Soc., London, Series A*, 236:71–73, 1958.

[16] T. Kayaba and K. Kato. The adhesive transfer of the slip-tongue and the wedge. *A S L E Trans.*, 24(2):164–174, 1981.

[17] T.E. Fischer, M.P. Anderson, and S. Jahanmir. Influence of fracture toughness on the wear resistance of yttria-doped zirconium oxide. *J. Am. Ceram. Soc.*, 72(2):252–257, 1989.

[18] J.M. Challen, P.L.B. Oxley, and B.S. Hockenhull. Prediction of archard's wear coefficient for metallic sliding friction assuming a low cycle fatigue wear mechanism. *Wear*, 111:275–288, 1986.

[19] M. Godet. The third body approach: A mechanical view of wear. *Wear*, 100:437–452, 1984.

[20] I. V. Kragelski and M.N. Dobycin. *Grundlagen der Berechnung von Reibung und Verschleiß*. VEB-Verlag Technik, Berlin, 1. Auflage edition, 1982.

[21] G. Fleischer, H. Bosse, and u. a. Berechnung der Reibung auf energetischer Grundlage. *Wissenschaftliche Zeitung der TU Magdeburg*, 26(6):17–27, 1982.

[22] D. Shakhvorostov, K. Pöhlmann, and M. Scherge. An energetic approach to friction, wear and temperature. *Wear*, 257:124–130, 2003.

[23] M. Scherge and M. Dienwiebel. Notes from lectures tribologie a-b. Technical report, Karlsruhe Institute of Technology, Karlsruhe, Germany, 2008-2009.

[24] D.M. Pirro and A.A. Wessol. *Lubrication Fundamentals*. Exxon Mobil, second edition, revised and expanded edition, 2001.

[25] R. Stribeck. Die wesentlichen Eigenschaften der Gleit-und-Rollen-Lager. *Z. VDI*, 46:1341–1348, 1902.

[26] C. Neumann. Tribologische Untersuchungen dentaler Oberflächen. Master's thesis, Fraunhofer Institut für Werkstoffmechanik (IWM) and Technische Universität Ilmenau, Technische Physik I / Oberflächenphysik, 2009.

[27] M. Scherge, K. Pöhlmann, and A. Gervé. Wear measurement using radio-nuclide-technique (rnt). *Wear*, 254:801–817, 2003.

[28] M. Moseler, P. Gumbsch, C. Casiraghi, A. C. Ferrari, and J. Robertson. The ultrasmoothness of diamond-like carbon surfaces. *Science*, 309(5740):1545–1548, 2005.

[29] I. Szlufarska, M. Chandross, and R. W Carpick. Recent advances in single-asperity nanotribology. *J. Phys. D: Appl. Phys.*, 41(12):123001 (39pp), 2008.

[30] M. Eriksson, J. Lord, and S. Jacobson. Wear and contact conditions of brake pads: dynamical in situ studies of pad on glass. *Wear*, 249(3-4):272 – 278, 2001.

[31] S. M. Rubinstein, G. Cohen, and J. Fineberg. Contact area measurements reveal loading-history dependence of static friction. *Phys. Rev. Lett.*, 96(25), 2006.

[32] W. Gregory Sawyer and Kathryn J. Wahl. Accessing inaccessible interfaces: In situ approaches to materials tribology. *MRS Bulletin*, 33:1145 – 1150, December 2008.

[33] T. Zijlstra, J. A. Heimberg, E. van der Drift, D. Glastra van Loon, M. Dienwiebel, L. E. M. de Groot, and J. W. M. Frenken. Fabrication of a novel scanning probe device for quantitative nanotribology. *Sensor. Actuat. A: Phys.*, 84(1-2):18–24, 2000.

[34] Nam P. Suh. The delamination theory of wear. *Wear*, 25(1):111–124, 7 1973.

[35] J. J. Pamies-Teixeira, N. Saka, and N. P. Suh. Wear of copper-based solid solutions. *Wear*, 44(1):65–75, 8 1977.

[36] D. A. Rigney. Comments on the sliding wear of metals. *Tribol. Int.*, 30(5):361–367, 5 1997.

[37] D. A. Rigney, L. H. Chen, M. G. S. Naylor, and A. R. Rosenfield. Wear processes in sliding systems. *Wear*, 100(1-3):195–219, 12 1984.

[38] J. P. Hirth and D. A. Rigney. Crystal plasticity and the delamination theory of wear. *Wear*, 39(1):133–141, 8 1976.

[39] P. Heilmann, J. Don, T. C. Sun, D. A. Rigney, and W. A. Glaeser. Sliding wear and transfer. *Wear*, 91(2):171–190, 11 1983.

[40] K.J. Wahl, R.R. Chromik, and G.Y. Lee. Quantitative in situ measurement of transfer film thickness by a newton's rings method. *Wear*, 264(7-8):731 – 736, 2008.

[41] M. Yoshino, T. Aoki, T. Shirakashi, and R. Komanduri. Some experiments on the scratching of silicon:: In situ scratching inside an sem and scratching under high external hydrostatic pressures. *Int. J. Mech. Sci.*, 43(2):335 – 347, 2001.

[42] Y.S. Zhang, Z. Han, K. Wang, and K. Lu. Friction and wear behaviors of nanocrystalline surface layer of pure copper. *Wear*, 260(9-10):942 – 948, 2006.

[43] A. Moshkovich, V. Perfilyev, I. Lapsker, and L. Rapoport. Stribeck curve under friction of copper samples in the steady friction state. *Tribol. Lett.*, 37:645–653, 2010.

[44] A. Moshkovich, V. Perfilyev, T. Bendikov, I. Lapsker, H. Cohen, and L. Rapoport. Structural evolution in copper layers during sliding under different lubricant conditions. *Acta Mater.*, 58(14):4685 – 4692, 2010.

[45] H. Kato, M. Sasase, and N. Suiya. Friction-induced ultra-fine and nanocrystalline structures on metal surfaces in dry sliding. *Tribol. Int.*, 43(5-6):925 – 928, 2010.

[46] L. Balogh, T. Ungar, Y. Zhao, Y.T. Zhu, Z. Horita, C. Xu, and T.G. Langdon. Influence of stacking-fault energy on microstructural characteristics of ultrafine-grain copper and copper-zinc alloys. *Acta Mater.*, 56(4):809 – 820, 2008.

[47] A. Mishra, B.K. Kad, F. Gregori, and M.A. Meyers. Microstructural evolution in copper subjected to severe plastic deformation: Experiments and analysis. *Acta Mater.*, 55(1):13 – 28, 2007.

[48] A. Moshkovich, V. Perfilyev, I. Lapsker, D. Gorni, and L. Rapoport. The effect of grain size on stribeck curve and microstructure of copper under friction in the steady friction state. *Tribol. Lett.*, 42(10):89–98, April 2011.

[49] R. Schouwenaars, V.H. Jacobo, and A. Ortiz. Microstructural aspects of wear in soft tribological alloys. *Wear*, 263(1-6):727 – 735, 2007.

[50] N.M. Alexeyev. On the motion of material in the border layer in solid state friction. *Wear*, 139(1):33 – 48, 1990.

[51] V. Panin, A. Kolubaev, S. Tarasov, and V. Popov. Subsurface layer formation during sliding friction. *Wear*, 249(10-11):860 – 867, 2001.

[52] G. Salomon, A.W.J. De Gee, and J.H. Zaat. Mechano-chemical factors in mos2-film lubrication. *Wear*, 7(1):87 – 101, 1964.

[53] S. Jahanmir and N. P. Suh. Mechanics of subsurface void nucleation in delamination wear. *Wear*, 44(1):17–38, 8 1977.

[54] Z. Liu, J. Sun, and W. Shen. Study of plowing and friction at the surfaces of plastic deformed metals. *Tribol. Int.*, 35(8):511 – 522, 2002.

[55] J.-D. Kamminga and G.C.A.M. Janssen. Experimental discrimination of plowing friction and shear friction. *Tribol. Lett.*, 25:149–152, 2007.

[56] L.G. Hector and S.R. Schmid. Simulation of asperity plowing in an atomic force microscope part 1: Experimental and theoretical methods. *Wear*, 215(1-2):247 – 256, 1998.

[57] K. Komvopoulos, N. Saka, and N.P. Suh. Plowing friction in dry and lubricated metal sliding. *J. Tribol.*, 108:301 – 312, 1986.

[58] K. Komvopoulos, N.P. Suh, and N. Saka. Wear of boundary-lubricated metal surfaces. *Wear*, 107:107 – 132, 1986.

[59] L. Wang, Y. He, J. Zhou, and J. Duszczyk. Modelling of plowing and shear friction coefficients during high-temperature ball-on-disc tests. *Tribol. Int.*, 42(1):15 – 22, 2009.

[60] P. Gilormini and E. Felder. Theoretical and experimental study of the ploughing of a rigid-plastic semi-infinite body by a rigid pyramidal indenter. *Wear*, 88(2):195 – 206, 1983.

A. Structure of LabView Software Implementation for the Tribometer

P.T.O.

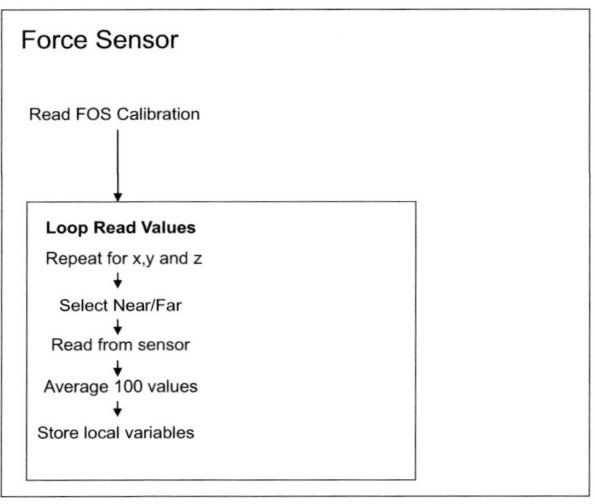

Force Sensor
Read FOS Calibration
Loop Read Values
Repeat for x,y and z
Select Near/Far
Read from sensor
Average 100 values
Store local variables

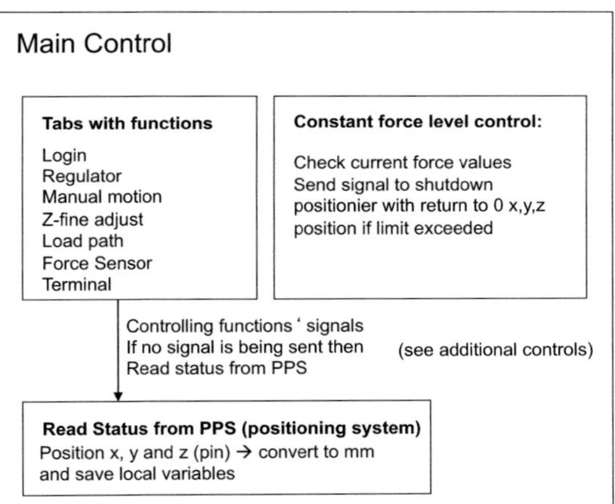

Main Control
Tabs with functions
Login
Regulator
Manual motion
Z-fine adjust
Load path
Force Sensor
Terminal
Constant force level control:
Check current force values
Send signal to shutdown
positionier with return to 0 x,y,z
position if limit exceeded
Controlling functions ' signals
If no signal is being sent then
Read status from PPS
(see additional controls)
Read Status from PPS (positioning system)
Position x, y and z (pin) → convert to mm
and save local variables

Additional Controls

Define current activity:
True/false variables trigger functions

If all triggers are false, the current status is updated within every cycle

Monitoring

Display elements:
Communication port
Current position x, y and z (pin)
Current force x,y and z
Pin height z on axis

Switch:
Enable/Disable force regulation

Graphical representation	
Current position	**Microscope image**
Draw and save grid in memory Convert current position from status to pixel position Place dot for current xy position Update image with new positions → continuous line.	Import and display last acquired image

Microscope Communication

Transfer protocol:

Use com channel for microscope → Finalize command → Transfer data to microscope PC

↑ Prepare and format command string

↓ Receive response / data packages

Tabs

Login

PPS (positioning system):
Initialize communication
Log on to PPS subsystem via TCP
Transfer Reference values for Z-axis

Microscope:
Initialize communication → receive confirmation response + key
Promt for password
Run internal script with encoding key → transmit password

Com established with confirmation code

Initialize positioning system

Select function
(Startup, Shutdown, Shutdown without returning to 0 position,
confirm centralization of positionier, complete startup with
activation of power supply, startup without reference motion)

Transmit message to positioner

Receive, decrypt and display response

Tabs

Manual positioning

Move to set value:
Set position: if pps status=available → transmit new coordinates
for x and y or z with predefined speed
Incremental motion: read current position, define new position
by adding distance multiplied with speed factor → transmit new
position with predefined speed (from speed factor)

Pin approach disabled if preset contact normal force has been
reached

Z fine positioning

Read z force value
Read current z position
Calculate difference from defined normal force for experiment
(this also defines the direction of motion)
Read set speed and distance factor for motion of z axis
Calculate new z position and speed
Transmit data to positioner

Set path

Load path:
Open excel file conaining path data
Convert to array
Compile command with path array
Upload data to positioner

Path related commands:
Execute path
Execute path loop
Stop at next defined position
Immediate stop
Resume
Resume in loop
→ Send command to positioner and read response

Tabs

Force sensor

Currently measured voltage:
Graphical representation of voltage for x,y,z sensors (within a
preset time limit)
Tare force:
Save current distance as local variable
Current force:
Calculate and display force acording to deviation from tare-value
Measured force:
Graphical representation of force for x,y,z sensors (within a
preset time limit)

Data storage

Select path to save data
Select data to store (position x,y,z, forces x,y,z and current path
cycle if loop is active)
Sample rate defines the frequency of data storing
Data entries are directly saved to a file with line feed
Definition of directory for microcope images (this is incorporated
in the string of the commands)

Terminal

Contains a constantly updated string field that presents
important messages received when calling up the status of the
positioning system.

Performing experiment

START:
Change status to performing experiment
Enter loop after checking if status of experiment is active

1. Write name for first microscope image
Using the path defined in storage and the current cycle of the
experiment, the name of the file is inserted in the command text
2. Send commands to Microscope
Acquire an image in double wavelength mode
Pause 1500 ms
Save file to defined location
Pause 4000 ms
3. Path
Execute single pass of defined path
4. Update cycle number
Add a cycle to local variable for current cycle

STOP:
Change status to experiment not active

List of Figures

List of Tables

Schriftenreihe
des Instituts für Angewandte Materialien

ISSN 2192-9963

Die Bände sind unter www.ksp.kit.edu als PDF frei verfügbar oder
als Druckausgabe bestellbar.

Band 1   Prachai Norajitra
           Divertor Development for a Future Fusion Power Plant. 2011
           ISBN 978-3-86644-738-7

Band 2   Jürgen Prokop
           **Entwicklung von Spritzgießsonderverfahren zur Herstellung
           von Mikrobauteilen durch galvanische Replikation.** 2011
           ISBN 978-3-86644-755-4

Band 3   Theo Fett
           **New contributions to R-curves and bridging stresses –
           Applications of weight functions.** 2012
           ISBN 978-3-86644-836-0

Band 4   Jérôme Acker
           **Einfluss des Alkali/Niob-Verhältnisses und der Kupferdotierung
           auf das Sinterverhalten, die Strukturbildung und die Mikro-
           struktur von bleifreier Piezokeramik $(K_{0,5}Na_{0,5})NbO_3$.** 2012
           ISBN 978-3-86644-867-4

Band 5   Holger Schwaab
           **Nichtlineare Modellierung von Ferroelektrika unter
           Berücksichtigung der elektrischen Leitfähigkeit.** 2012
           ISBN 978-3-86644-869-8

Band 6   Christian Dethloff
           **Modeling of Helium Bubble Nucleation and Growth
           in Neutron Irradiated RAFM Steels.** 2012
           ISBN 978-3-86644-901-5

Band 7   Jens Reiser
           **Duktilisierung von Wolfram. Synthese, Analyse und Charak-
           terisierung von Wolframlaminaten aus Wolframfolie.** 2012
           ISBN 978-3-86644-902-2

Band 8   Andreas Sedlmayr
           **Experimental Investigations of Deformation Pathways
           in Nanowires.** 2012
           ISBN 978-3-86644-905-3

Band 9   Matthias Friedrich Funk
            **Microstructural stability of nanostructured fcc metals during
            cyclic deformation and fatigue.** 2012
            ISBN 978-3-86644-918-3

Band 10   Maximilian Schwenk
            **Entwicklung und Validierung eines numerischen Simulationsmodells
            zur Beschreibung der induktiven Ein- und Zweifrequenzrandschicht-
            härtung am Beispiel von vergütetem 42CrMo4.** 2012
            ISBN 978-3-86644-929-9

Band 11   Matthias Merzkirch
            **Verformungs- und Schädigungsverhalten der verbundstranggepressten,
            federstahldrahtverstärkten Aluminiumlegierung EN AW-6082.** 2012
            ISBN 978-3-86644-933-6

Band 12   Thilo Hammers
            **Wärmebehandlung und Recken von verbundstranggepressten
            Luftfahrtprofilen.** 2013
            ISBN 978-3-86644-947-3

Band 13   Jochen Lohmiller
            **Investigation of deformation mechanisms in nanocrystalline
            metals and alloys by in situ synchrotron X-ray diffraction.** 2013
            ISBN 978-3-86644-962-6

Band 14   Simone Schreijäg
            **Microstructure and Mechanical Behavior of Deep Drawing DC04 Steel
            at Different Length Scales.** 2013
            ISBN 978-3-86644-967-1

Band 15   Zhiming Chen
            Modelling the plastic deformation of iron. 2013
            ISBN 978-3-86644-968-8

Band 16   Abdullah Fatih Çetinel
            **Oberflächendefektausheilung und Festigkeitssteigerung von nieder-
            druckspritzgegossenen Mikrobiegebalken aus Zirkoniumdioxid.** 2013
            ISBN 978-3-86644-976-3

Band 17   Thomas Weber
            **Entwicklung und Optimierung von gradierten Wolfram/
            EUROFER97-Verbindungen für Divertorkomponenten.** 2013
            ISBN 978-3-86644-993-0

Band 18   Melanie Senn
            **Optimale Prozessführung mit merkmalsbasierter
            Zustandsverfolgung.** 2013
            ISBN 978-3-7315-0004-9